Analysis and Management of Productivity and Efficiency in Production Systems for Goods and Services

Analysis and Management of Productivity and Efficiency in Production Systems for Goods and Services

Fabio Sartori Piran, Daniel Pacheco Lacerda,
and Luis Felipe Riehs Camargo

CRC Press
Taylor & Francis Group
Boca Raton London New York

CRC Press is an imprint of the
Taylor & Francis Group, an **informa** business

This book was previously published in 2018 in Portuguese by Elsevier Editora, Brazil, as *Análise e Gestão da Eficiência*, by Luis Felipe Riehs Camargo, Daniel Pacheco Lacerda and Fabio Antonio Sartori Piran.

CRC Press
Taylor & Francis Group
6000 Broken Sound Parkway NW, Suite 300
Boca Raton, FL 33487-2742

First issued in paperback 2021

© 2020 by Taylor & Francis Group, LLC
CRC Press is an imprint of Taylor & Francis Group, an Informa business

No claim to original U.S. Government works

ISBN-13: 978-0-367-35772-6 (hbk)
ISBN-13: 978-1-03-217577-5 (pbk)
DOI: 10.1201/9780429351679

Library of Congress Cataloging-in-Publication Data

Names: Camargo, Luis Felipe Riehs, author. | Piran, Fabio Sartori, author. | Lacerda, Daniel Pacheco, author.
Title: Analysis and management of productivity and efficiency in production systems for goods and services / by Fabio Sartori Piran, Daniel Pacheco Lacerda, Luis Felipe Riehs Camargo.
Other titles: Análise e gestão da eficiência. English
Description: Boca Raton, FL : CRC Press/Taylor & Francis Group, 2020. | "This book was previously published in 2018 in Portuguese by Elsevier Editora, Brazil, as Análise e gestão da eficiência by Luis Felipe Riehs Camargo, Daniel Pacheco Lacerda and Fabio Antonio Sartori Piran." | Includes bibliographical references and index.
Identifiers: LCCN 2019042478 (print) | LCCN 2019042479 (ebook) | ISBN 9780367357726 (hardback ; acid-free paper) | ISBN 9780429351679 (ebook)
Subjects: LCSH: Industrial efficiency. | Production management.
Classification: LCC T58.8 .C3613 2020 (print) | LCC T58.8 (ebook) | DDC 658.5--dc23
LC record available at https://lccn.loc.gov/2019042478
LC ebook record available at https://lccn.loc.gov/2019042479

Visit the Taylor & Francis Web site at
http://www.taylorandfrancis.com

and the CRC Press Web site at
http://www.crcpress.com

Contents

Foreword

Describing how an organization behaves in the face of competition and what its objectives are is a fundamental task for any institution that intends to perpetuate in the market. Likewise, it is critical for a nation to know how it is developing, and whether its policies and plans for growth and improvement of the population's living conditions are effective. But how to do this? What are the key indicators and measurements to identify how well an organization, institution or country is doing? How can we tell if a company is managing to produce better than it did before and/or better than its competition? How can we know if a country's government policies are effective and achieving the desired goals and objectives? How can we ascertain that the economic development plans are taking effect?

Among the indicators and metrics used to measure the aforementioned factors, productivity and efficiency are fundamental, as much in the strictest context of companies and organizations as in a broader context, involving a state or a whole country. However, the concepts of productivity and efficiency, at times confused, may lead organizations to take wrong decisions due to mistakes in the interpretation and use of the results obtained. Such confusion in terms of concepts may be a consequence of the flexibility and diversity of measurements that can be created for both productivity and efficiency. Since productivity is defined as the ratio of inputs to outputs, the factors to be considered in the numerator and denominator of this equation may be the most diverse. Similarly, when we calculate the efficiency, this is always based on the ratio of what is actual output and what could (should) be accomplished. Again, considering the numerator and denominator of this equation, these may be the most varied factors. This flexibility in the factors that can be considered in the productivity and efficiency equations, as well as the similarity of the equations that calculate them, namely a ratio between two factors, makes the concepts of productivity and efficiency often confused and misused.

Maybe a simple way to differentiate them is to note that efficiency always compares equal units of measurement, that is, how much was done compared with how much could have been done, or the actual time taken to produce compared with the time that could have been taken. As the numerator and denominator have the same unit of measurement, efficiency has no unit and is expressed in percentage terms. The same does not happen with productivity, because it is calculated based on inputs and outputs that have different units of measurement. Thus, productivity could be measured, for example, in parts/hour or tons/kWh^{-1}.

In the context presented, where we need to measure the productivity and efficiency of public and private organizations and institutions, the difficulty in understanding the concepts of productivity and efficiency and the confusion between them, and the empirically proven need for techniques and tools to measure these concepts, makes the purpose of this book of the utmost importance for professionals faced with the need to take decisions based on the efficiency and productivity of the organization, as well as for students training for careers.

It should be noted that, in the case of this book, the authors focus their attention on the most technical aspects of productivity and efficiency, that is, the analysis of these measures in the context of organizations. To this end, as an introductory part of the book, important concepts, definitions and classifications about productivity and efficiency are covered, serving as a bibliographic source for professionals and students who seek a better understanding of this topic. The simple, precise form of writing in this chapter can be observed throughout the book, making it easy and pleasant to read.

The book, in addition to providing clarifications regarding the concepts covered, presents case studies and applications of efficiency and productivity. This allows the reader to identify with the situations and readily understand the content. Furthermore, the book addresses current concepts and proposals, fruits of a modeling research group focused on Production Engineering at GMAP/UNISINOS. Thus, there are chapters that present contributions developed during cutting-edge research in the area, and which have resulted in Master's dissertations and Doctoral theses.

It is also worth highlighting the focus on Data Envelopment Analysis (DEA), a technique for Operational Research, widely used for the calculation of efficiency, and which can also make many contributions to the Production Engineering area. This technique, although widely used in the scientific community and presenting several characteristics with potential to contribute to enterprise, is still hardly utilized in companies.

Seeking to aid expansion of the use of DEA in the reality of organizations, this book presents a modeling method for DEA application. This method, due to its ease of application and the simple explanation of its steps in Chapter 6, could represent a turning point for increased tool utilization in the business environment.

Finally, given that the main part of the book is the content and not this Foreword, I would like to make a brief comment and a compliment about how it came about, and, hopefully, how many others will be realized. The emergence of the book itself, of course, is the result of discussions among the authors about some works that are currently being carried out. However, what should be emphasized is the process by which this group of authors joined forces to produce this work. The process involved the creation of a serious, multi-disciplinary, multi-ideological and multi-functional research group. The group had, and still has, different views of how this techno-scientific work in Production Engineering should be performed. When reading the finished book, it is difficult to imagine and appreciate all the work performed prior to its production. Therefore, I would like to congratulate the authors for working so seriously and professionally toward the evolution of Production Engineering in the country, making several contributions, including this book.

Finally, I would like to wish you all productive reading and efficient use of the learning to be gained.

Ricardo Augusto Cassel, PhD
Professor at Universidade Federal do Rio Grande do Sul (UFRGS).

A Solutions Manual, Exercises, Instructor's Manual, PowerPoint slides, and figure slides are available for download at the book's website. Please visit https://www.crcpress.com/Analysis-and-Management-of-Productivity-and-Efficiency-in-Production-Systems/Piran-Lacerda-Camargo/p/book/9780367357726.

Acknowledgments

Fabio Sartori Piran

Publishing this book is for me a great personal achievement. I am always happy to contribute in some way to businesses, as well as to academia and the public in general. I believe this book is a way to make this contribution. This work was made possible due to the highly collaborative partnership I enjoy with the other authors. I thank Prof. Dr. Daniel Lacerda, for whose work and character I have a lot of respect and admiration. Daniel has been my tutor, advisor and academic referee. My life has changed for the better since we started working together on our research. In fact, we share great enthusiasm for research in Operations Management, and I believe we will go much further together on this journey. I also thank Prof. Dr. Luis Felipe, who has helped me a great deal in understanding the technique we present in this book. Luis Felipe was my Master's degree co-advisor, and we are good friends. As with Daniel, I also believe that we will go far together on this journey. Although this work was written by me, Daniel and Luis Felipe, there are other researchers who also contributed. Special thanks go to friends, Rafael Marques, Roberta Bondan, Charles Von Gilsa, Iberê Guarani de Souza and Alaércio De Paris. I am very grateful to them for the excellent work they carried out, and these are duly reported in our book. Thanks go to all the GMAP/UNISINOS colleagues for their help and encouragement. Also to friends, Maria Isabel Wolf Motta Morandi, Aline Dresch, Douglas Veit, Pedro Lima, Carolina Araujo, Jaqueline Abreu e Débora Nunes and the other members of this great group. Special thanks are due to my friend, Luis Henrique Rodrigues. I am very grateful to Flávio I. Kubota and Paulo A. Cauchick-Miguel, who accepted the invitation to write a chapter and help us in the construction of the work. My special thanks go to my parents, João Piran and Lidia Sartori Piran, who were and still are my best "teachers", instilling in me values such as honesty, ethics and a sense of justice. I am grateful to my first teacher, Terezinha Piran (my Dinda), who has always been an exemplary teacher, in my opinion. Finally, I express special gratitude to my wife, Camila Timm Neves for her support and affection.

Daniel Pacheco Lacerda

I started my scientific and research journey exactly 15 years ago, when my Master's degree began. In 2019, together with my two important companions, I concluded this, my 15th book in the area of Management and Production Engineering. Therefore, first and foremost, it is appropriate to begin by thanking my two partners in this work. Thanks are due to Prof. Dr. Luis Felipe Riehs Camargo for his collaboration since 2013 in the supervision of several dissertations in the Graduate Program in Production and Systems Engineering - PPGEPS/UNISINOS. This fruitful partnership resulted in two ABEPRO (Brazilian Association of Production Engineering) awards for best dissertation in Brazil. Both dissertations used the concepts and techniques covered in this book. Above all, Luis Felipe has been a great partner in academic discussions on the most diverse subjects. Our convergence of thought makes many people say that we are brothers. From this point of view, they are not wrong. I think we particularly look like this when we play FIFA. My life trajectory made it possible to get to know Prof. Fabio Sartori Piran, with whom I have been on a productive, enjoyable academic journey. Fabio is one of the people with whom I share many values: ethics, love of work and scientific rigor. Deepest thanks are due to him for his partnership in all the challenges we have set ourselves. I hope that destiny will bring us ever closer, as we undoubtedly have much to build on. Thanks are due to all GMAP/UNISINOS members (Learning Modeling Research Group - http://gmap. unisinos.br). Deepest thanks go to Prof. Pedro Lima, my current fellows, and those who were part of the research group for the entire partnership, especially Prof. Dr. Douglas Rafael Veit and Prof. Dr. Maria Isabel Wolf Motta Morandi. Gratitude is due to them for their valuable partnership, and for jointly valuing aspects that are considered fundamental, such as truth, seriousness, work, self-awareness, self-respect and desire to continually improve. Many talk about these values, but few exercise them in practice. Even at the most difficult times in GMAP/UNISINOS, we have been drawn ever closer. Thanks are also due to my other great companions on this journey: Prof. Dr. André Riberio (UERJ), Prof. Dr. Renato Cameira (Poli/UFRJ), Prof. Dr. Édison Renato (Poli/COPPE/UFRJ), Prof. Rafael Barbaestefano (CEFET/ Rio), Prof. Dr. Rafael Teixeira (University College of Charleston, USA), Prof. Dr. Paulo Cauchick (UFSC), Prof. Dr. Carlo Bellini (UFPB), Prof. Dr. Lucila Campos (UFSC), Prof. Rafael Paim (CEFET/Rio), Prof. Dr. Liane Kipper (UNISC), Prof. Dr. Julio Siluk (UFSM), Prof. Dr. Adriano Proença (Poli/UFRJ) and Prof. Dr. Heitor Caulliraux (Poli/COPPE/UFRJ). I also extend my gratitude to my eternal source of inspiration, Prof. Dr. Ione Bentz (PPGDesign/UNISINOS); I cannot imagine what my life would have been like without her. Thanks to Prof. Dr. Ricardo Augusto Cassel (UFRGS), my former teacher, a friend of many years, as well as part of my family more recently. My admiration for him includes his wholeheartedness, professionalism, thoughtfulness and education (I would like him to know that I mirror him very much in the raising of my children). Many thanks are due to Prof. Dr. Junico Antunes for all his fellowship (it has been 15 years). Thanks to Dr. Luis Henrique Rodrigues, great friend and partner. His statements and critical points of view have greatly improved this book. Luis will forever be in my heart. Thanks to Prof. Aline Dresch, who has been a great friend, and a partner many times in recent years. The moments we have had together are unforgettable, productive and pleasurable. I see in her a lot of the qualities I value in people, not to mention her potential as a great professional. I hoped others will also see and value her qualities, and work toward her success. This is the least she deserves, and I would like her to always remember that "the criterion of truth is practice". May life's decisions and paths bring us ever closer. Thanks to one of my great intellectual partners, Dr. Priscila Ferraz, who started in 2006, for all her conversations, understanding, partnership and mutual learning. Being around her has been a unique experience. May life never separate us. I have admired her for a long time and she has never disappointed. Finally, I would like to thank all my family: my wife, Carina (Xuxu), for her support and affection in all matters that have involved us. God has blessed my wife and I for our greatest "projects" (loves): Caio Lacerda and Serena Lacerda, who are better than we could ever have asked for. I love their affectionate hugs at times of tiredness, and even for the "chicken" (as Serena calls the mess she has made in the whole house) that actually makes me feel so happy. I am also so grateful to God for everything and for all the wonderful people who have come and gone in my life. As the great Gabriel Garcia Marquez would say: "Remembering is easy for those who have memories, forgetting is difficult for those with hearts." My heart will NEVER forget them!

Luis Felipe Riehs Camargo

This book contemplates the materialization of a long and productive partnership with professors and researchers, Daniel Lacerda and Fabio Sartori Piran. It was 2012 when I returned from my sandwich doctorate at Erasmus University (Rotterdam) and was invited by my good friend Daniel Lacerda to assist with the mentoring of Charles. He was a Master's student in production engineering, who was working with the DEA technique to evaluate production efficiency at a petrochemical plant in Rio Grande do Sul State. I soon understood the magnitude and potential of DEA to generate essential information to support the decision-making process in organizations. Since then, I have been involved in many works, developed both in the academic sphere and published in Master's dissertations and scientific articles, as well as in work carried out in important companies in the national scenario. That said, I thank Daniel Lacerda deeply for his friendship, long conversations and experience exchange, for the opportunity to delve into DEA research and applications, and for his invitation to participate in this book project. I thank Prof. Guilherme Liberali, who not only opened the door for the sandwich doctorate, but also provided entry into the world of analytics and quantitative modeling. Additionally, I thank Prof. Fabio Sartori Piran, who understood and fully immersed himself in DEA studies, culminating in the development of this book. Thanks are due to my friends and colleagues at GMAP/UNISINOS for providing a healthy, productive space in which to move forward with the research and DEA applications, especially Prof. Luis Henrique Rodrigues, a critic of the DEA technique who provided alternative ways of strengthening DEA, such as in the mechanisms of variable selection and conceptual modeling. I wish I could name all those who have crossed my path and made my life better, but this is risky as I could leave someone out. Therefore, I express my gratitude to all my family, friends, teachers/professors, colleagues, students and clients. I want everyone who reads this book to be grateful to them. Last, but by no means least, in addition to God, I express my special thanks to my closest partner, my beloved wife, Laura Trapp, who has made my life infinitely better over the past 16 years.

Authors

Fabio Sartori Piran: PhD candidate in Industrial and Systems Engineering at Universidade do Vale do Rio dos Sinos, UNISINOS. Master's degree in Industrial and Systems Engineering at UNISINOS. Bachelor's degree in Logistics from UNISINOS. Higher Education in Production Management at Universidade FEEVALE. Researcher at GMAP/UNISINOS. Consultant partner in the Hahn Piran Costs Business, a company providing consultancy and training in the areas of operations management, production systems, process mapping, cost management and control. Experience in costs management with emphasis on cost reduction, loss analysis, control and organization of inventory, and implementation of management tools. Also works in the industrial area with emphasis on planning, programming and production and materials control (PPCPM), chrono analysis, layout and production management. Has developed major projects in small, medium and large national and multinational companies in the fields of footwear, artifacts, textiles, engineering, food and retail. Professor at Universidade FEEVALE.

Daniel Pacheco Lacerda: PhD in Production Engineering awarded by COPPE/ UFRJ. Master's degree in Administration from UNISINOS (2005). CNPq Research Productivity Scholar (PQ-2) in the Production Engineering area since 2017. A co-ordinator of the Bachelor Program in Production Engineering, in UNISINOS. Possesses professional and academic experience in the areas of operations strategy, business process engineering, data envelopment analysis, design science research and restriction theory. Currently, a permanent researcher in the Graduate Program in Industrial and Systems Engineering, PPGEPS/UNISINOS, and an academic co-ordinator of the GMAP/UNISINOS (Research Group on Modeling for Learning). Develops applied research projects in companies and public entities, such as FIOCRUZ/Bio-Manguinhos, PETROBRAS, TRANSPETRO, JBS, AGDI, SEBRAE/RS, SESI and VALE. Awarded academic honors: Outstanding Paper Award for Excellence of the Emerald Literati Network, orientation of dissertations awarded by ABEPRO in 2013/2014/2015, Gaúcho Researcher Award, FAPERGS 2014 (category: Researcher in Company), and a CNPq grant to study Productivity in Innovation, Technological Development and Extension (2013–2016).

Luis Felipe Riehs Camargo: PhD in Administration from UNISINOS (2013) with a period spent at Erasmus University Rotterdam (Supervisor: Guilherme Liberali Neto). Earned a Master's degree in Production Engineering from UNISINOS (2009) and a Bachelor's degree in Electrical Engineering from Pontifícia Universidade Católica do Rio Grande do Sul (2005). Experience as a researcher in computer simulation projects and mathematical modeling in projects developed at UNISINOS and Erasmus University. Also develops applied research projects in collaboration with companies and other organizations, such as Samarco, Petrobras, SEBRAE, Vale and Grupo Randon.

1 Introduction

Productivity and efficiency are widely debated concepts, whether in academia or among business and government professionals. Despite the number of studies on this theme, Brazil has been facing difficulties in distinguishing the two concepts. According to the magazine *Exame* (2017a), between 1981 and 1990, the productivity per worker in the country fell by an average of 2% a year, whereas, from 1991 to 2000, it rose on average by 1.6% a year. This trend remained positive between 2001 and 2010, increasing at an annual rate of 1.2%. In the period 2011–2016, productivity per worker reverted to a negative path, presenting a mean annual decline of 1.1%. The data contrast with those from the period 1950–1980, when work productivity grew on average by 3.5% per year. It is to be noted that this was the principal factor responsible for the significant growth in income per capita in the country (averaging 3.9% a year) throughout the period. However, as of the 1980s, productivity stagnated. Among the salient factors, one may cite the reduced adoption of new technology and taxation distortions that led to inefficient allocation of funds (Exame, 2017a).

Among the reasons explaining the low Brazilian productivity that lasted for decades, there were structural factors related to: (i) technology (in today's production structure, sectors show a low capacity to incorporate technology and produce better products); (ii) poor educational standards; and (iii) inadequate occupational skills training. Deficient infrastructure is also recognized as a significant factor, added to which excessive bureaucracy, complex taxation and an unfavorable business environment explain the scenario of production stagnation (EXAME, 2017b).

Productivity and efficiency constitute factors of extreme importance in companies' competitiveness, for economic sectors and entire countries. Productivity, as much as efficiency, may be observed on diverse scales of analysis, ranging from the macro level of whole economies to the micro scale in the ambit of sectors, processes and the companies themselves. Besides this, distinct factors interfere with and explain productivity and efficiency in accordance with the level of analysis desired.

What hinders the researchers' and professionals' understanding is that productivity and efficiency may be considered in both technical and economic terms. In general, the analyses realized for economic sectors and whole countries are economic. In turn, in organizations, the analysis is in technical terms with the broad use of Overall Operational Performance Index (OOPI)/ Overall Equipment Effectiveness (OEE). As will be seen in subsequent chapters, the primacy of OOPI/OEE may lead to limitations in the analytical capacity, and, as a consequence, reduce managers' range of options for action.

Thus, our focus is directed toward analysis of productivity and efficiency in the company context. Such a focus does not exclude use of the concepts, techniques and tools presented for broader contexts, such as economic sectors and entire countries. In addition, productivity and economic efficiency may also be analyzed, even if only partially, based on what will be described in later chapters.

In this book, we seek to bolster the analysis of productivity and efficiency by means of Data Envelopment Analysis (DEA). DEA is a broadly divulgated technique and is used by the scientific community for a range of applications. However, its use in everyday life and the business environment is developing. It is, therefore, an important instrument to complement and qualify analyses that are performed centrally by OOPI/OEE. Among the benefits of DEA use are the following:

i. Simultaneous analysis of multiple outputs and inputs in a single efficiency measurement.
ii. Identification of the benchmarks over the periods of analysis.
iii. Analysis of productivity and efficiency in constant or variable economies of scale.
iv. Analysis oriented toward maximization of outputs or reduction of inputs in an operation, process or organization.
v. Analysis and definition of the goals and opportunities for improvement based on identification of targets and slack calculation.
vi. Support for assessment of external and internal benchmarks.

Although this book presents a mathematical structure that is rather advanced for organizations, we have sought to structure it in such a way as to facilitate its use via different mechanisms. First, we seek to present a solid definition of the concepts that permeate the phenomena of productivity and efficiency. Second, we present the metrics and techniques used to operationalize and measure these concepts. Third, we describe DEA in depth and a method for its operationalization in organizations. Fourth, we develop a computational platform to enable use of the examples provided, and for realization of analyses with the use of DEA in companies. Finally, evidence is provided from case studies of its application, originating from the scientific research conducted over the past 10 years. Some of these cases were published in the main scientific journals in the operations management area (for example, *International Journal of Production Economics, Benchmarking* and *International Journal of Advanced Manufacturing Technology*).

Thus, this book has arisen from a line of research and set of investigations conducted in GMAP I UNISINOS (http://gmap.unisinos.br) since its establishment. This research focus has been manifested previously in recent discussions about the need to increase productivity and efficiency in Brazil, in a range of spheres of interest. Looking to the future, we can confidently state that the analysis of productivity and efficiency will remain, despite the need to incorporate and deepen the economic analysis.

Recent advanced manufacturing technology, for example, requires more profound analysis of the effects on production systems in technical and economic terms. The simple use of Augmented Reality, Additive Manufacturing, Cyber-Physical Systems and Automation, among others, are not justified per se. These technologies need to raise the productivity and efficiency of the production systems.

One additional aspect to be considered refers to the need to improve the productivity and economic efficiency of organizations. Considering the high costs of acquisition, implementation and operationalization of the technologies associated with

Advanced Manufacturing, additional attention to this aspect is necessary. Therefore, although there may be an increase in productivity and efficiency in technical terms, it would imply a reduction in economic terms. As a consequence, a company should not lose sight of its goal, "Earn money today and in the future" as Eliyahu Goldratt affirms in his novel, *The Goal*. In other words, in the final analysis, all the efforts employed in the technologies of Advanced Manufacturing should revert to an increase in productivity and economic efficiency in organizations. In terms of Advanced Manufacturing, we seek to illustrate Modularity, which is a technology that could contribute to an increase in productivity and efficiency.

Finally, we have sought to cover an important gap in production and operations management in the Brazilian literature. The analysis and management of productivity and efficiency, although treated sparingly and with a focus on the tools to increase them, is lacking in terms of concepts, techniques and tools for the analysis itself. There is a presupposition that use of OOPI/OEE is sufficient for companies. However, as may be observed, this metric, albeit necessary, is not sufficient for a broad profound analysis and management of productivity and efficiency in organizations.

REFERENCES

Exame (2017a). Acesso em 31 de janeiro de 2018. Disponível em: https://exame.abril.com.br/economia/produtividade-brasileira-nao-cresce-desde-1980-diz-estudo/. Acesso em: 31 de janeiro de 2018.

Exame (2017b). Acesso em: 31 de janeiro de 2018. Disponível em: https://exame.abril.com.br/revista-exame/para-crescer-de-verdade-brasil-precisa-vencer-atraso-de-50-anos/.

Goldratt, E. M., & Cox, J. (2002). *A meta: um processo de melhoria contínua* (2nd ed.). São Paulo: Nobel.

2 Vision of Productivity and Efficiency from a Systemic Perspective

This chapter presents the concepts related to productivity and efficiency. Understanding the difference between productivity and efficiency is important, as, despite being distinct concepts, in many cases they are treated as if they were the same. We deal with productivity and efficiency from a systemic perspective, analyzing the system as a whole (the production of goods as much as the provision of services), avoiding distortions derived from an assessment that considers the whole as the sum of the parts. In addition, in this chapter we present other important concepts, such as efficacy, effectiveness and benchmarking.

Improvement in productivity and efficiency is a contemporary challenge for organizations that produce goods and services. As a consequence, it is necessary to measure productivity and efficiency precisely, objectively and as a whole. Analysis of productivity and efficiency allows managers to qualify their decision-making. Such decisions could lead to better use of resources, cost reduction, better allocation of investment and more precise definition of goals, among others.

Although productivity and efficiency are distinct concepts, in many cases, as mentioned above, they are treated as synonyms. In this book, we will differentiate them in the following manner. *Productivity* is the relation between the input resources and the results, the outputs generated by machinery, an operation, a process or a system (Charnes et al., 1978), that is, the ratio of outputs to inputs. *Efficiency* is a comparative measurement that represents the exploitation of resources, that is, what was produced with the use of certain resources compared to what could have been produced with the same resources (Cummins & Weiss, 2013).

However, the productivity index is related to the efficiency index. If a productivity index of a goods production or service unit was compared with the productivity index of a unit with better performance, a relation would be formed in which it would be possible to make a comparison between these units. This comparative relation is the efficiency index, in which the more productive unit is used as a reference (Førsund, 2017a).

In many companies, efficiency is measured by calculation of the ratio between the hours worked and the hours available for production. This calculation refers only to the control of operational efficiency and is limited by not considering a set of other resources used in the goods and services production processes (for example, materials, indirect labor, general production expenditure, etc.). Such limitations restrict use of information about efficiency for managerial decision-making in organizations. It is not a recent development that efficiency should be perceived as a matter that considers the entire company (overall analysis), and not only the labor efficiency (partial analysis of the system) (Skinner, 1974). Therefore, utilizing an inadequate efficiency measurement may be insufficient for assessment of companies' performance.

Furthermore, in the majority of companies, the forms of assessment of productivity and efficiency are specific and local, that is, only parts of the system are assessed. There is no method that favors the assessment of productivity and efficiency from a broader systemic perspective, one that allows specific actions with a focus on the overall increase in operational performance. The lack of an overall (systemic) analysis of productivity and efficiency can lead an organization to taking wrong decisions, such as: (i) unnecessary investment in resources of lower priority; (ii) lack of investment in critical resources; and (iii) investment in increased productive capacity without exploring the restrictions (bottlenecks) and the maximum capacity of the existing resources.

Faced with these aspects, a metric and a method that systematically assesses overall productivity and efficiency of the production systems are necessary, as much in academic terms as in practical ones. Presentation of this metric and the proposition of this method are among the objectives of this book. In order to develop the proposed method, we sought to use a systemic vision, which criticizes reductionism through an understanding generated by phenomena upon dividing them into parts, and then conducting the study of these, simply observing in terms of cause and effect (Morandi et al., 2014). Systemic thinking possesses a characteristic, namely a vision, based on the indivisible whole. In opposition to mechanistic thinking, which seeks understanding of a system through its parts, systemic thinking has, as its unit of analysis, the system itself (Andrade et al., 2006).

2.1 PRODUCTIVITY IN A PRODUCTION SYSTEM OF GOODS AND SERVICES

Productivity, in a broad context, provides evidence of how a particular production process uses its resources. Productivity refers to the volume of items produced with a certain quantity of inputs, that is, relates what is produced to what is consumed (Cummins & Weiss, 2013). Productivity may be understood to be the relation between the quantity of goods or services generated (outputs) and the quantity of resources consumed in their generation (inputs) in the same time interval (t) (Heizer & Render, 2001), in accordance with Equation 2.1.

$$\text{Productivity} = \frac{\text{Outputs}}{\text{Inputs}} \quad (2.1)$$

Productivity may vary according to the differences in the production technologies available to organizations, in the execution of the operation plan, in the very sequence of operations in a certain process, and in the environment in which the production occurs. Analysis of these factors leads to identification of alternatives that can enable increased productivity (Ferreira & Gomes, 2009).

Specifically, in the area of services, adoption of the traditional productivity concept can lead to companies having a false perception of the performance of their operations (Durdyev et al., 2014). This is because the majority of the existing productivity measurement models are derived from the industrial context (Armistead & Machin, 1998).

The operations in manufacturing take place without direct participation of the customers, and quality is a determining factor based on the technical specifications of the products. In services, most operations occur in a transparent interactive system, where the client is able to intervene in the productivity and quality (Grönroos & Ojasalo, 2004).

The active client participation in the service process limits standardization of the operations and activities, and, consequently, that of the variables to be used for productivity assessment. This lack of standardization of the variables hinders definition of the quantity of *inputs* necessary to realize a given *output*. From this perspective, the interactions of the service providers with their clientele may influence productivity in the process of rendering the service (Johnston & Jones, 2004).

In short, measurement of productivity in service operations differs due to the special characteristics inherent in services. Among the principal characteristics, the following are outstanding: client participation in the process, intangibility, heterogeneity, inseparability, perceptibility and time (Djellal & Gallouj, 2013; Fitzsimmons & Fitzsimmons, 2014).

For example, the variable relative to time indicates that the results of a service operation cannot be considered immediately after its provision. They should also be examined within a certain period after the service has been concluded (Djellal & Gallouj, 2013). In order to exemplify the time characteristic, one may consider a hospital service. The immediate result is associated with a change in the patient's state of health soon after the service has been rendered, whereas the long-term result is associated with alterations in the patient's state in the period after consuming the service. The very characteristics of the customers, besides their participation in the process, impact productivity.

From this perspective, the concept of productivity created for a manufacturer requires adaptations for application to services. Vuorinen et al. (1998) highlighted that the models used to measure productivity in services should consider quantitative and qualitative aspects. Chart 2.1 presents the main differences between service and manufacturing companies in measuring productivity.

Following discussion of the concepts of productivity, the concepts regarding efficiency are presented below.

2.2 EFFICIENCY

The studies on efficiency conducted by Debreu (1951) and Farrell (1957) are considered remarkable, as they affirm that the efficiency measurements in manufacturing and service companies are fundamental factors for conception of theories and suitable policies. Farrell (1957) commented that the techniques known for this measurement at that time were not capable of improving performance. Farrell (1957) attributed this fault to the fact that industry considered as a performance factor only the relationship between the production and the work (labor) employed for such. Thus, various other resources fundamental for technical and economic performance

CHART 2.1

Differences in Measuring Productivity in Services and Manufacturing

Manufacturing	Services
Production is first, consumption later	Production and consumption are simultaneous
Productivity is measured in a closed system	Productivity is measured in an open system
The quality depends on the result	The quality depends on the result and the process
Inputs and outputs are clear	It is difficult to differentiate between inputs and outputs
Customers do not participate in the production process	Customers participate in the production process
Customers do not generate inputs directly for the production process	Inputs achieved by the customers affect process performance
Products can be stocked	Services cannot be stocked

Source: Marques (2017).

were ignored. In addition to being ignored, these other resources could actually limit labor use itself.

Farrell's proposal (1957) was directed at the way companies could use the inputs of their production processes to transform them into outputs. With this, they could seek tification of the consumption of these inputs in relation to the production that they would obtain in their factories. A company becomes technically efficient when it uses a minimal volume of inputs to obtain a particular volume of products (Cummins & Weiss, 2013). In this way, the greater the number of results obtained for a certain quantity of resources, the higher the efficiency. Coelli et al. (2005) highlighted that efficiency in producing goods and providing services can be analyzed from two aspects, technical efficiency and scale efficiency. In addition, there is also allocative efficiency and cost/economic efficiency.

2.2.1 TECHNICAL AND SCALE EFFICIENCY

Technical efficiency is related to the capacity of a process to produce a certain quantity of products or services, utilizing the least amount of inputs in relation to the other processes observed (Cummins & Weiss, 2013). Technical efficiency can also be understood as a skill to obtain maximum production from a particular set of inputs. On the one hand, it becomes possible to produce a greater volume of results with the use of the same resources. On the other hand, it becomes possible to produce the same results utilizing a lower volume of resources (Farrell, 1957).

Efficiency of scale is the result of the level of maximum production situated under the efficient frontier. Thus, points off the optimal scale of production are not fully efficient (Färe et al., 1994). Scale efficiency consists of an optimal functioning unit where a reduction or increase in the scale of production implies reduced efficiency. Scale efficiency is related to the use of the best production practices. Identification of the productivity and efficiency of each unit of analysis allows the recognition of which units are more efficient and can be used as reference (benchmarking) for the others.

2.2.2 ALLOCATIVE AND COST/ECONOMIC EFFICIENCY

The models of allocative efficiency should be preferred when the data of the variables (inputs and outputs) used are measured and represented by monetary units (Portela, 2014). Allocative efficiency reflects the capacity to minimize costs, utilizing inputs in optimal proportions, taking into account their prices (Ferreira & Gomes, 2009).

Allocative efficiency should be interpreted as the proportion of costs of a decision-making unit in relation to the minimum cost observed to produce a certain level of results (Portela, 2014). In other words, allocative efficiency is the ratio between the unit cost to the company to produce its products and the unit cost to produce them using the best production practice (Førsund, 2017a). The outcome of the combination of technical and allocative efficiency is known as cost/economic efficiency (Portela, 2014), as displayed in Figure 2.1.

Cost/economic efficiency is a broader concept than technical and allocative efficiency, since it also involves optimized choice of the levels and mix of inputs and/or outputs based on market price behavior. To be efficient in costs, the company has to define its levels of input and/or output to optimize a certain economic objective, generally the minimization of costs or maximization of profit (Bauer et al., 1998). For this, the cost/economic efficiency requires a combination of technical and allocative efficiency. Thus, cost/economic efficiency (CE) is calculated by the multiplication of technical efficiency (TE) by allocative efficiency (AE), (CE = TE * AE).

It is quite plausible for some technically efficient companies to be economically inefficient and vice versa. This depends on the relation between the managers' skills to use resources better and their skills to respond to market signs. Therefore, the use of different concepts of efficiency may provide different rankings of the companies (Bauer et al., 1998; Coelli et al., 2005). In this manner, when information about costs and prices are available, in addition to analysis of technical efficiency, economic efficiency is recommended.

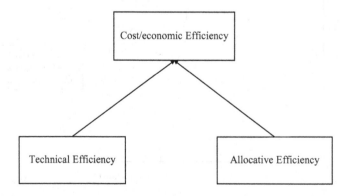

FIGURE 2.1   Illustration of cost/economic efficiency.

2.3 EFFICIENCY VS. PRODUCTIVITY

The terms productivity and efficiency are frequently used in companies and in the literature (Coelli et al., 2005; Cummins & Weiss, 2013; Bartelsman et al., 2013; Para Coelli et al., 2005). They are often treated as synonyms, but, as presented previously, they do not have the same definitions. Equation 2.2 aims at helping to reinforce this understanding.

$$\text{Productivity} = \frac{\text{Outputs}}{\text{Inputs}}$$

$$\text{Efficiency} = \frac{\text{Outputs}}{\text{Inputs}} \text{ realized and COMPARED with maximum} \frac{\text{Outputs}}{\text{Inputs}} \quad (2.2)$$

If a company only analyzed productivity, it is possible that the resources would not be properly used (Bartelsman et al., 2013). To illustrate the difference between efficiency and productivity, Coelli et al. (2005) exemplified this with a simple production process with a sole input (X) to produce just one output (Y). Figure 2.2 represents such a process, facilitating understanding of the distinction between inputs (inputs: x axis) and outputs (outputs: y axis).

The production frontier (Curve F) represents achievement of the maximum possible output in relation to each input (Coelli et al., 2005). If the company analyzed in Figure 2.2 operates at point A, it can be considered inefficient, as it is technically possible to increase production to the level of point B without a need to increase the inputs, as represented by the finest line.

The companies situated at points B and C can be considered efficient, as they are on the curve, which represents the efficiency frontier. The company represented by letter C is the most productive, as it lies in the region of the curve in which productivity is at the maximum possible. Company A is the only one outside the efficient

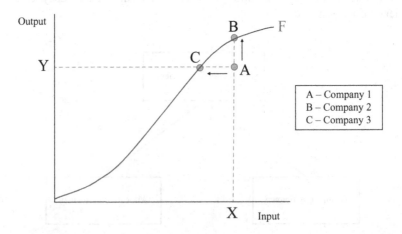

FIGURE 2.2   Frontiers of production and technical efficiency.

frontier, which leads to the conclusion that this company is not efficient and, at the same time, not productive.

In order to understand the analysis of productivity and efficiency from the systemic perspective, it is also important to understand other concepts. Therefore, in the next sections, concepts and the relations among efficacy, effectiveness and benchmarking will be discussed.

2.4 EFFICACY AND EFFECTIVENESS

Assessment of efficacy consists of analysis of the achievement of the goals established for a particular project, for example. In this case, the monitoring of efficacy, which occurs at the end of the project, presents information about the goals achieved and whether they correspond to or diverge from what was proposed at the start of the project (Minayo, 2011). Conceptually, efficacy represents fulfilment of the objectives proposed, without considering the resources used (Førsund, 2017b). It can be said that something is efficacious if the final objective has been achieved, irrespective of the resources used.

Assessment of effectiveness captures the effects of an improvement project or program, and has the purpose of measuring the quantitative and qualitative changes caused by an intervention (Minayo, 2011). Thus, effectiveness considers the "before" and "after" of the execution of a change or compares the results obtained in another situation under similar conditions, in which there has been no change. In addition, effectiveness may be related to the capacity to achieve the objectives proposed, considering the resources used (Førsund, 2017b). This includes the choice of the objectives and the appropriate methods to achieve them. Chart 2.2 presents a synthesis of the concepts addressed up to this point in the chapter.

2.5 BENCHMARKING

Benchmarking originated in the Xerox Corporation in 1979 with the objective of identifying which company branch presented the best operational performance. After this identification, it became possible to verify which practices accounted for this high performance and thus replicate them in the other branches.

Benchmarking is defined as the quest for a company's best practices that lead to higher performance (Camp, 1989; Vinodh & Aravindraj, 2015). It can also be understood as a continuous systematic process used to measure and compare the business process of an organization in relation to the leading enterprises in any part of the world to obtain information that aid the taking of measures aimed at improving its performance (Rakesh et al., 2008). Through benchmarking, one can identify the processes, practices and managerial methods that provide better results, and adapt them to the organization of interest.

Furthermore, benchmarking can serve as support in seeking continuous improvement. It can be applied to production processes. However, it is important that the units compared be similar, in terms of the inputs and the production result (outputs). Carpinetti and De Melo (2002) pointed out that benchmarking may be classified according to Chart 2.3.

CHART 2.2
Synthesis of Concepts

Concept	Description
Productivity	Relation between resources of inputs used and results (outputs) generated by a machine, an operation, a process or a system
Efficiency	Comparative measurement that represents the use of the resources, that is, the outputs produced with use of certain resources compared to what could have been produced with the same resources
Technical efficiency	Capacity of a process to produce a particular quantity of products or services, utilizing the least number of inputs in relation to the processes observed
Efficiency of scale	Result of the level of maximum production situated under the efficient frontier, which consists of an optimal unit of operation where the reduction or increase in the scale of production implies a reduction in efficiency
Allocative efficiency	Capacity to minimize the costs, utilizing the inputs in optimal proportions, taking the prices of these inputs into account
Cost/economic efficiency	Combination of technical efficiency with allocative efficiency
Efficacy	Capacity to achieve the objectives proposed, without considering the resources used
Effectiveness	Capacity to achieve the objectives proposed, considering the resources used

CHART 2.3
Classification of Benchmarking

Type	Description
Process benchmarking	Used to compare operations, work practices and business processes
Product benchmarking	Used to compare products and/or services
Strategic benchmarking	Used to compare organizational structures, management practices and business strategies

Besides this, benchmarking may be external or internal. Section 2.5.1 discusses internal benchmarking concepts, and 2.5.2 deals with those of external benchmarking.

2.5.1 INTERNAL BENCHMARKING

Internal benchmarking constitutes an important mechanism to be used in companies in which an external benchmark is difficult or even inviable (Southard & Parente, 2007). Internal benchmarking may be considered a comparison of a company's different production units. Besides this, a comparison may be made of one production unit with itself, that is, considering its performance over time. In these cases, for example, one can seek to understand within the firm itself why one production or service unit presents better results than the others. After identification, the best practices may be replicated in the other units, thereby increasing the organization's profitability as a whole (Bogetoft & Otto, 2010; Southard & Patente, 2007). The comparison may also be made with similar products or services in similar business units (Vinodh & Aravindraj, 2015; Cook et al., 2017).

2.5.2 EXTERNAL BENCHMARKING

In external benchmarking, a comparison is made of the company's own performance with those of competitors. Normally, visits are paid to other enterprises to get to know, analyze and perhaps adopt practices identified. It can be affirmed that this is the form of benchmarking most commonly used by organizations. External benchmarking may be subdivided into three categories (Carpinetti & Melo, 2002), as described in Chart 2.4.

2.5.3 HOW TO APPLY BENCHMARKING

Figure 2.3 shows, in general terms, a step-by-step definition, developed by Carpinetti and De Melo (2002), of the object of study, benchmarking in companies.

In Step 1, information should be gathered about the characteristics of the product, the customer/client and the target market. In addition, one may verify information regarding competitive priorities and fabrication and financial strategies, among other areas. This will aid understanding about which variables and activities are crucial to improvement in competitiveness.

In Step 2, information should be collected about the customers' expectations and the quality for different categories of customers and products. Afterward, there should be classification of the relative importance of the requirements for the key customers. Also, information needs to be garnered about performance in catering for client expectations in comparison to that of competitors. This helps to identify the dimensions with possibilities of improvement.

CHART 2.4

Categories of External Benchmarking

Category	Description
Benchmarking of competition	Used to compare a company's performance with that of a direct competitor with the same product. In this case, a comparison may be made of products, services or business processes.
Functional benchmarking	Used to compare specific functions with the best practices. It is a benchmarking process application that compares a specific function in two or more organizations in the same industry.
Generic benchmarking	Used to seek the best practices independently of the industry. It is similar to functional benchmarking, but the objective is to compare with the best in the field, without taking the industry into account.

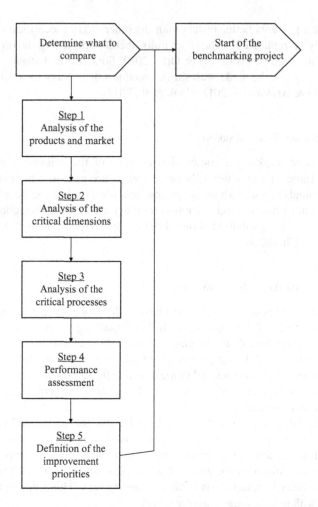

FIGURE 2.3   Steps to define the object of study, benchmarking. (Adapted from Carpinetti and De Melo, 2002.)

In Step 3, mapping should be made of all the processes and activities that are relevant to or that support the added-value chain. In addition, it is important to understand the relationship with the dimensions that offer improvement opportunities. This may be assisted by means of the construction of a matrix that relates the processes with the dimensions. In turn, this would help to focus attention on the processes and activities that most affect performance in the priority competitive dimensions.

In Step 4, a qualitative or quantitative assessment should be made of the performance in critical processes and activities. Diagnosis of the current situation is fundamental to understanding which areas or activities are weak points that need to be addressed. The quantitative information, if available, may reveal areas and dimensions that need attention.

In Step 5, it is understood that, after performing the analysis proposed in Steps 1–4, the dimensions and activities that require improvement should become evident.

From this point on, the improvement project itself may begin to deal with matters for which a comparative assessment would be considered appropriate.

One of the criticisms of the application of benchmarking is that normally the information collected from the processes used as reference is qualitative. Even when the information is quantitative, it becomes difficult to establish improvement goals with the data obtained. Overcoming this difficulty is one of the objectives of the technique and the content that will be presented in this book. Besides this, the concepts, techniques and tools presented may allow the realization of benchmark objectives, even in situations where there are no comparable units externally or internally.

2.6 RELATION AMONG THE CONCEPTS

Figure 2.4 presents the relations among the concepts presented. Initially, it is necessary to identify the most productive unit, the one with the best capacity to transform inputs into outputs. The most productive unit may be considered the most efficient unit, which may be compared with the others that are inefficient (benchmarking).

However, the most efficient unit may not be the most effective, because the most effective production unit is the one that executes the best practices, but, for example, it may not achieve the proposed production goal. In order to use the resources in an efficient manner and also meet the production goals, the production unit needs to be considered effective. The effective productive unit may also be considered as a benchmark in the analysis carried out.

The technique explored in this book considers the relations between productivity and efficiency. Although a general understanding of these topics is important, the concepts of efficacy and effectiveness are not considered in the analysis we present.

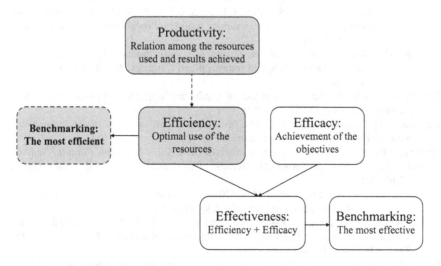

FIGURE 2.4   Relation among the concepts.

REFERENCES

Andrade, A. L., Seleme, A., Rodrigues, L. H., & Souto, R. (2006). *Pensamento Sistêmico Caderno de Campo: o desafio da mudança sustentada nas organizações e na sociedade*. Porto Alegre: Bookman.

Armistead, C., & Machin, S. (1998). Business process management: Implications for productivity in multi-stage service networks. *International Journal of Service Industry Management*, 9(4), 323–336.

Bartelsman, E., Haltiwanger, J., & Scarpetta, S. (2013). Cross-country differences in productivity: The role of allocation and selection. *The American Economic Review, 103*(1), 305–334.

Bauer, P. W., Berger, A. N., Ferrier, G. D., & Humphrey, D. B. (1998). Consistency conditions for regulatory analysis of financial institutions: A comparison of frontier efficiency methods. *Journal of Economics and Business, 50*(2), 85–114.

Bogetoft, P., & Otto, L. (2010). *Benchmarking with Dea, Sfa, and R (Vol. 157)*. Springer Science & Business Media.

Camp, R. C. (1989). Benchmarking: The search for best practices that lead to superior performance. *Quality Progress, 22*(1), 61–68.

Carpinetti, L. C., & De Melo, A. M. (2002). What to benchmark? A systematic approach and cases. *Benchmarking: An International Journal, 9*(3), 244–255.

Charnes, A., Cooper, W. W., & Rhodes, E. (1978). Measuring the efficiency of decision-making units. *European Journal of Operational Research, 2*(6), 429–444.

Coelli, T. J., Rao, D. S. P., O'Donnell, C. J., & Battese, G. E. (2005). *An introduction to efficiency and productivity analysis* (2nd ed., pp. 349). New York: Springer.

Cook, W. D., Ruiz, J. L., Sirvent, I., & Zhu, J. (2017). Within-group common benchmarking using DEA. *European Journal of Operational Research, 256*(3), 901–910.

Cummins, J. D., & Weiss, M. A. (2013). Analyzing firm performance in the insurance industry using frontier efficiency and productivity methods. In *Handbook of insurance* (pp. 795–861). New York: Springer.

Debreu, G. (1951). The coefficient of resource utilization. *Econometrica: Journal of the Econometric Society, 19*(3), 273–292.

Djellal, F., & Gallouj, F. (2013). The productivity challenge in services: Measurement and strategic perspectives. *The Service Industries Journal, 33*(34), 282–299.

Durdyev, S., Ihtiyar, A., Ismail, S., Ahmad, F. S., & Bakar, N. A. (2014). Productivity and service quality: Factors affecting in service industry. *Procedia-Social and Behavioral Sciences, 109*, 487–491.

Färe, R., Grosskopf, S., Norris, M., & Zhang, Z. (1994). Productivity growth, technical progress, and efficiency change in industrialized countries. *The American Economic Review, 84*(1), 66–83.

Farrell, M. J. (1957). The measurement of productive efficiency. *Journal of the Royal Statistical Society, 120*(3), 253–290.

Ferreira, C. M. C., & Gomes, A. P. (2009). *Introdução à análise envoltória de dados: teoria, modelos e aplicações*. Viçosa: Editora UFV.

Førsund, F. R. (2017a). Economic interpretations of DEA. *Socio-Economic Planning Sciences*.

Førsund, F. R. (2017b). Measuring effectiveness of production in the public sector. *Omega, 73*, 93–103.

Fitzsimmons, J. A., & Fitzsimmons, M. J. (2014). *Administração de Serviços-: Operações, Estratégia e Tecnologia da Informação*. AMGH Editora.

Grönroos, C., & Ojasalo, K. (2004). Service productivity: Towards a conceptualization of the transformation of inputs into economic results in services. *Journal of Business Research, 57*(4), 414–423.

Heizer, J. H., & Render, B. (2001). *Administração de operações: bens e serviços*. LTC.

Johnston, R., & Jones, P. (2004). Service productivity: Towards understanding the relationship between operational and customer productivity. *International Journal of Productivity and Performance Management, 53*(3), 201–213.

Marques, R. (2017). *Análise da eficiência e das influências do cliente em operações de serviços* (Dissertação de mestrado em Engenharia de Produção e Sistemas pela Universidade do Vale do Rio dos Sinos – Unisinos).

Minayo, M. C. (2011). Importância da avaliação qualitativa combinada com outras modalidades de Avaliação. *Saúde & Transformação Social, 2*(2), 02–11.

Morandi, M. I. W. M., Rodrigues, L. H., Lacerda, D. P., & Pergher, I. (2014). Foreseeing iron ore prices using system thinking and scenario planning. *Systemic Practice and Action Research, 27*(3), 287–306.

Portela, M. C. A. S. (2014). Value and quantity data in economic and technical efficiency measurement. *Economics Letters, 124*(1), 108–112.

Rakesh et al. (2008). The propagation of benchmarking concepts in Indian manufacturing industry. *Benchmarking: An International Journal, 15*(1), 101–117.

Skinner, W. (1974). The focused factory. *Harvard Business Review, 52*, 113–121.

Southard, P. B., & Parente, D. H. (2007). A model for internal benchmarking: When and how? *Benchmarking: An International Journal, 14*(2), 161–171.

Vinodh, S., & Aravindraj, S. (2015). Benchmarking agility assessment approaches: A case study. *Benchmarking: An International Journal, 22*(1), 2–17.

Vuorinen, I., Järvinen, R., & Lehtinen, U. (1998). Content and measurement of productivity in the service sector: A conceptual analysis with an illustrative case from the insurance business. *International Journal of Service Industry Management, 9*(4), 377–396.

3 Practical Problems in Relation to Productivity and Efficiency

Following a discussion of the concepts of productivity and efficiency from a systemic perspective, in this chapter we examine practical problems that are part of the quotidian of organizations that possess goods production systems or service provision systems, and may be analyzed with the DEA technique. Afterwards, in the next chapters, DEA will be analyzed, so that, with a better understanding of DEA, we will demonstrate in Chapter 10 how the technique can contribute to analysis, management and solution of these problems.

3.1 EFFICIENCY AND INTERVENTION IN THE SYSTEM: CASE STUDY IN A BUS MANUFACTURER

Improvement in efficiency is a contemporary challenge for industrial organizations, and implementation of improvement programs are considered to be strategic alternatives to achieving these objectives. Among the improvement programs, one may consider the modularization of products, which can be regarded as a design strategy to manufacture modular products that allow greater variety at lower cost, rather than designing exclusive products for each market segment or specific client requirements (Salvador et al., 2002).

Modularization makes it viable to increase the variety of products, as it is presupposed there is planning, development and production of components with the capacity to generate combinations that allow the formation of a wide range of final products (Starr, 1965; Starr, 2010). The concept of modularization of products is described in greater depth in Chapter 11.

An increase in productivity and efficiency is expected from product modularization (Starr, 1965; Starr, 2010; Patel & Jayaram, 2014). However, in order to confirm the effects of the use of product modularization, it is necessary to measure them. This may be considered a practical problem related to the measurement of productivity and efficiency in a goods production system.

To contribute to this approach, a case study was conducted to assess whether product modularization really provides an improvement in productivity and efficiency. This study occurred in a bus manufacturer. The market for bus manufacturers is characterized by personalization of products, due to the need to cater for customers' different requirements. Therefore, the vehicles may vary in terms of size, number of seats, baggage space and optional accessories, among others. Aiming at the viability of increasing the variety of products offered to the market without prejudicing operational performance, in 2007 the company began the development and fabrication of vehicles with modular architecture.

Initially, the vehicles were designed and produced with six large modules: (i) front; (ii) rear; (iii) right side; (iv) left side; (v) roof; and (vi) base. Subsequently, the modules of the two sides were subdivided into smaller modules. Besides this, modules were developed for the interior, such as the instrument panel, partitions, seats and the electrical system.

Measurement was realized in two areas of the company, Product Engineering and the Production Process. The analysis aimed at understanding if modularization had effects on the efficiency of Product Engineering and the Production Process. For this, the efficiency of these areas was assessed over time, contemplating the periods before and after modularization.

To investigate the effect of modularization, two types of product projects (Product Engineering) and two types of final products (Production Process) were selected, one modularized and the other, non-modularized. In Product Engineering, the project of the non-modularized product (nMP) was considered to be the control variable, where the project had not undergone intervention (modularization), the effects of which were being investigated. The project of the modularized product (MP) was considered to be the response variable, where the project was subjected to intervention (modularization). In the Production Process, the same procedures were carried out, the non-modularized product (nMP) being the control variable, and the modularized product (MP), the response variable.

It is understood that simultaneously observing these projects and products (modularized and non-modularized) is a necessary condition to better observe the effects of modularization on productivity and efficiency. Besides this, the reliability of the results obtained increases, as it facilitates attribution of the results observed to the intervention under study. After understanding the technique presented for analysis of productivity and efficiency, in Chapter 10, Section 10.1, we will show how this problem can be analyzed.

3.2   EFFICIENCY IN SERVICES: CASE STUDY IN AN AUTOMOTIVE FLEET MANAGEMENT COMPANY

Low productivity and inefficiency in service companies generate recurring discussions, and present challenges in the operations area. Besides this, in many cases, productivity and efficiency are measured incorrectly by service company management (Djellal & Gallouj, 2013). Therefore, a study was conducted aiming to contribute to assessment of productivity and efficiency in the services area in an automotive fleet management company.

The company where the study was conducted is a nationwide vehicle hire operator. The company assesses the productivity of its operation by means of the ratio between the operational revenue generated by the services and the sum of the costs and expenses of the operation. The current way of assessing productivity and efficiency presents limitations, as it does not consider various resources consumed during the service generation process. Consideration of the greatest amount of consumed resources possible is a fundamental rule for analyses of productivity and efficiency.

Furthermore, the relationship between the operational revenue and the costs and expenses is an aggregate indicator, which precludes measurement of efficiency by means of comparison of the productivity of different services. The limitations of the productivity indicator, and the absence of an indicator to measure efficiency, impose difficulties on decision-making with respect to resource management, cost reduction and definition of goals and objectives for the organization.

It was also found that the ratio between the operational revenue and the costs and expenses is an indicator that does not assess the productivity of each client contract, and, as a consequence, does not demonstrate which contracts are efficient. Thus, it is not possible to define with precision the quantity of resources necessary to render the contracted services.

The company in which the research was conducted provides services to customers that wish to outsource the management of their vehicle fleet. As stated above, the company provides vehicle rental services, fleet rotation, control of documentation and fines, management of fleet maintenance and other services.

The customers participate in the company's service provision process and are in a position to influence the planned results. However, the company does not assess the possible benefits the client interactions may generate in the operation. Besides this, alterations in the efficiency indexes of the services may depend on customers' actions, given their aforementioned participation (Djellal & Gallouj, 2013).

In order to illustrate the participation of the customers in the company process, one can study the fleet maintenance management service. This service will be productive by dealing with the occurrences in the fleet (service transactions) at the lowest possible cost. It is understood that variables under the client's control, such as the age of the fleet, degree of wear, and the make of the vehicles, may impact operation efficiency.

Besides not assessing the efficiency of its operation, the organization under study does not assess efficiency from its customers' perspective. Thus, it is not possible to identify whether the contracts are efficient for the customers, perhaps generating risk of dissatisfaction with the services. Such dissatisfaction may lead the company to lose contracts to competitors. Aiming to contribute to the analysis of productivity and efficiency in the company, an investigation was carried out seeking to analyze which variables in the service contracts may impact efficiency of the operation. In Chapter 10, Section 10.2, one can verify the results of this research.

3.3   EFFICIENCY AND BENCHMARKING: CASE STUDY IN A FUEL STATION NETWORK

For retail fuel stations, there is a need to increase productivity and efficiency due to a reduced profit margin, as the result is directly related to sales volume.

Focusing on competition in the market, the fuel stations need to identify the inefficient and efficient units in their operation, so that they may develop improvement actions and replicate the best practices. Analysis of efficiency may contribute to these aims, as it would enable managers to take decisions for overall improved performance in the operation (Hadi-Vencheh et al., 2014).

A discussion with the manager of a network of five fuel stations pointed out the need for an investigation to assess the same variables in another network of fuel stations, and to contribute to a management model of results for an increase in and continuous assessment of efficiency. Thus, this analysis had the objective of realizing an assessment over time of the efficiency of the units in a fuel station network. The results of this case study are presented in Section 10.3 of Chapter 10.

3.4   EFFICIENCY AND INVESTMENT PROJECTS: CASE STUDY OF A PETROCHEMICAL COMPANY

The companies in the petrochemical industry, faced with increased local and world competition, constantly seek consolidation of their competitive positions. The competition in this sector occurs because petrochemical products have limited differentiation, and, as a consequence, their prices are determined by the market. Thus, the results of the companies are basically defined by operational performance (Gilsa et al., 2017).

In this context, where an organization's performance basically depends on the efficiency of its operations, and the sale price is determined by the market, research was conducted to investigate whether or not the technological changes undertaken in the company's operations were improving their efficiency indexes. The research used DEA and the Malmquist Index jointly for assessment of the organization's productive efficiency in a longitudinal case study.

The research was conducted for one of the most important chemical companies in the world. The company develops, produces and sells special chemical products, plastic, rubber and intermediate products. It employs approximately 16,700 people in 30 countries and owns 47 production units all over the world.

The research was conducted at one of the production units situated in Brazil during the period 2004–2011. Initially, a calculation of the efficiency was made through DEA, utilizing the model with constant returns to scale (CRS), with input as the orientation. Thus, it was possible to determine the relative efficiency of each input, in order to identify which variable most influences efficiency of the DMU.

Subsequently, the efficiency of the model with a variable return to scale (VRS) was calculated with CRS to allow calculation of the efficiency of scale, which is the quotient of the efficiency with CRS, the result of division by the efficiency with VRS. Finally, it was sought to identify the DMUs that achieve increased or decreased efficiency.

The DMUs considered to be efficient were those with efficiency of scale equal to one (CRS), whereas those considered to be inefficient were those that presented lower efficiency values in the VRS model. In addition to this, the efficiency was segregated into three distinct periods. This was done to assess whether there was a significant difference between the means of efficiency among these periods. To complete the analysis, the Malmquist Index was calculated and used to identify displacement of the efficiency frontier among the periods of analysis.

Finally, statistical tests were used to analyze variance, such as the t, analysis of variance (ANOVA) and Kruskal-Wallis tests, with the objective of assessing whether any difference in efficiency found among the periods had statistical significance or not. Thus, it was possible to observe whether there was any real improvement in the company's efficiency over time, and if so, whether it could be attributed to technological upgrades in the company's process. The results of this case study are presented in Section 10.4 of Chapter 10.

3.5 EFFICIENCY AND CONTINUOUS IMPROVEMENT: CASE STUDY OF AN ARMS MANUFACTURER

Continuous improvement programs have been the object of study by researchers and managers in the operations management area. This occurs because these programs possess elements that make viable improvements in the performance of the process and use of the best practices in organizations. However, despite the benefits obtained, there are reports of a high failure rate in the employment of efforts to implement continuous improvement programs (Easton & Jarrell, 1998; Glover et al., 2011; McLean et al., 2015).

In addition to continuous improvement, organizations use learning curves to raise the operational performance and to understand the behavior of the processes (Terwiesch & Bohn, 2001; Franceschini & Galetto, 2003). The learning curve is useful to minimize the costs incurred in the introduction of a new product into the production process, among others. However, despite the benefits presented in relation to continuous improvement and the learning curve, a need becomes apparent for studies that measure the results obtained over time.

Studies normally focus only on an optimistic approach in relation to continuous improvement programs and learning. Therefore, this position does not represent the reality of many organizations. Thus, there is a risk of not knowing the true benefits of these programs.

With the analysis made, and with the aim of advancing in this direction, an investigation was conducted with an arms manufacturer. The company is situated in an old, obsolete industrial park, and the majority of its machinery and technology date back to the period 1930–1970. The principal measurement of efficiency used by the organization is the ratio between the hours available and the hours worked. In the company under analysis, there is no robust measurement for the assessment of efficiency, nor any measurement for assessment of the results of the continuous improvement programs and learning over time. Besides this, the company does not know the impact of the production volume on its efficiency, which may lead to investment in continuous improvement projects that are unnecessary.

The context presented reminds one of the difficulties in identifying the true inefficiency on production lines. With this, the company does not know the right moment to invest in new technology. The managers consider that the continuous improvement programs are mainly responsible for the company's development over time. Due to the antiquated industrial park, the company accumulates high costs in terms of maintenance, contracting consulting services, constant layout changes, tool breakages, and other expenses on programs focusing on continuous improvement.

Thus, the objective of this study was to realize an analysis of the relations among continuous improvement, learning, efficiency and production volume in the arms manufacturer. On the basis of the study, it is possible to verify whether or not there was improvement in the company's operational efficiency over time, considering the continuous improvement processes based on accumulation of knowledge and learning. The research was conducted by means of a longitudinal case study contemplating an investigation period of six years. As techniques of analysis, DEA, combined with ANOVA and linear regression analysis, were employed. The results of this case study are presented in Section 10.5 of Chapter 10.

REFERENCES

Djellal, F., & Gallouj, F. (2013). The productivity challenge in services: Measurement and strategic perspectives. *The Service Industries Journal*, *33*(34), 282–299.

Easton, G., & Jarrell, S. (1998). The effects of total quality management on corporate performance: An empirical investigation. *Journal of Business*, *71*(1), 253–307.

Franceschini, F., & Galetto, M. (2003). Composition laws for learning curves of industrial manufacturing process. *International Journal of Production Research*, *41*(7), 1431–1447.

Gilsa, C. V., Lacerda, D. P., Camargo, L. F. R., Souza, I. G., & Cassel, R. A. (2017). Longitudinal evaluation of efficiency in a petrochemical company. *Benchmarking: An International Journal*, *24*(7), 1786–1813.

Glover, W. J., Farris, J. A., Van Aken, E. M., & Doolen, T. L. (2011). Critical success factors for the sustainability of Kaizen event human resource outcomes: An empirical study. *International Journal of Production Economics*, *132*(2), 197–213.

Hadi-vencheh, A., Ghelejbeigi, Z., & Gholami, K. (2014). On the input/output reduction in efficiency measurement. *Measurement*, *50*, 244–249.

McLean, R. S., Antony, J., & Dahlgaard, J. J. (2015). Failure of continuous improvement initiatives in manufacturing environments: A systematic review of the evidence. *Total Quality Management & Business Excellence*, *28*(3), 219–237.

Patel, P. C., & Jayaram, J. (2014). The antecedents and consequences of product variety in new ventures: An empirical study. *Journal of Operations Management*, *32*(1–2), 34–50.

Salvador, F., Forza, C., & Rungtusanatham, M. (2002). Modularity, product variety, production volume, and component sourcing: Theorizing beyond generic prescriptions. *Journal of Operations Management*, *20*(5), 549–575.

Starr, M. K. (1965). Modular production: A new concept. *Harvard Business Review*, *3*, 131–142.

Starr, M. K. (2010). Modular production: A 45-year-old concept. *International Journal of Operation and Production Management*, *30*(1), 7–19.

Terwiesch, C., & Bohn, R. E. (2001). Learning and process improvement during production ramp-up. *International Journal of Production Economics*, *70*(1), 1–19.

4 Techniques for Calculation of Productivity and Efficiency

This chapter presents a summarized set of techniques for the calculation of productivity and efficiency. These techniques are Laspeyres Index (1864), Paasche Index (1874), Fisher Index (1922), Drobish Index, Törnqvist Index (1936), Free Disposal Hull (FDH), Thick Frontier Approach (TFA), Distribution-Free Approach (DFA), Analysis of Stochastic Frontier (SFA), Hierarchical Analysis Process (AHP), OEE (Overall Equipment Effectiveness) and Data Envelopment Analysis (DEA). Some of these techniques will be explained in more detail, others less so. Besides this, the potential applications and limitations of each technique will be discussed.

Improvement of the operational performance of manufacturing systems requires adoption of the best practices of production and methods, techniques and tools suitable for performance measurement (Kenyon et al., 2016). Performance measurement is a key process for companies, and one of the objectives of these measurement methods, techniques and tools is to provide analysis of the productivity and efficiency of organizations and production systems (Hitt et al., 2016; Balfaqih et al., 2016).

Coelli et al. (2005) argue that the calculation techniques of productivity and efficiency may be classified according to the nature of the data and the method. Regarding the nature of the data, it may be parametric or non-parametric. As for the method, it may be classified as frontier or non-frontier. The parametric methods are related to measurement of data that use an interval scale or a ratio, supported by parameters based on presuppositions for testing samples. The parametric methods also suppose a functional relationship and correlation between production and inputs. The non-parametric calculation methods are used when the parameters violate the presuppositions of the sample, that is, when the requirements in relation to the data used in the analysis are fewer (Ferreira & Gomes, 2009).

The non-frontier methods suggest that the maximum efficiency (1 or 100%) is known. Thus, the efficiency defined will be achieved by the unit under analysis, whose inputs may not be reduced and without there being a reduction in the quantity of its manufactured products, or quantity of which may not be increased without increasing inputs as well (Coelli et al., 2005).

For the frontier methods, maximum efficiency is achieved when one or more units under analysis attain performance higher than the others. The main efficiency and productivity calculation techniques and their respective classifications are synthesized in Chart 4.1.

CHART 4.1

Techniques for Efficiency and Productivity Calculations

	Non-Frontier	Frontier
Parametric	Laspeyre Index (1864)	The Frontier Approach (TFA)
	Paasche Index (1874)	Free Distribution Approach (DFA)
	Fisher Index (1922)	Stochastic Frontier Analysis (SFA)
	Drobish Index	
	Törnqvist Index (1936)	
	Free Disposal Hull (FDH)	
Non-Parametric	Hierarchical Process Analysis (AHP)	Data Envelopment Analysis (DEA)
	Overall Equipment Effectiveness (OEE)	

In the following sections, we will present the concepts regarding each technique for productivity and efficiency calculation, as shown in Chart 4.1. The objective is to provide a broad vision to assist with the choice of a suitable analysis technique, according to the particular context of investigation.

4.1 LASPEYRES INDEX (1864)

The Laspeyres Index (1864) adopts a fixed weighting base. Thus, it consists of a weighted arithmetic mean of the related indexes, the weighting factors being determined by the weights, the quantities and the inputs, in two periods, t and t+1 (Munneke & Slade, 2001). The relative indexes allow a comparison to be made between the two periods (t and t+1) for a single product. This analysis is expressed as a percentage, showing the changes that have occurred over time.

As an illustration, one may consider the price of a certain raw material today in relation to that paid the year before. In this case, the Laspeyres Index (1864) is calculated based on the ratio between the amount of money at current prices and that necessary to purchase the same raw material quantity at the base date, as shown in Equation 4.1. The amount of money at current prices is obtained through multiplication of the current price by the base quantity. The amount at the base date is obtained through multiplication of the base price by the base quantity.

$$\text{ILaspeyres} = \frac{\sum_{i=1}^{n} P_t^i * Q_0^i}{\sum_{i=1}^{n} P_0^i * Q_0^i} \tag{4.1}$$

where:

ILaspeyres = Laspeyre Index;

P_t = price of the good(s) in period t (current);

Q_0 = quantity of the good(s) at the base date;

P_0 = price of the good(s) at the base date.

4.2 PAASCHE INDEX (1874)

The Paasche Index (1874) adopts a mobile weighting base. Thus, it is considered to be an aggregate index, calculated by the mean of the indexes referring to quantities weighted by the relative importance of the products in the current period (Munneke & Slade, 2001). In this way, weights are considered for the inputs, calculated based on the prices and quantities of the goods in the current period, considering another distinct period (t and t+1), and taking as weights the quantities arbitrated for these inputs in the initial period.

The Paasche Index (1874) is obtained by calculation of the ratio between the amount of money for acquisition of a particular quantity of raw material at current prices and the amount of money necessary for the same acquisition at the base date (Equation 4.2). The amount of money at current prices is obtained through multiplication of the current price by the current quantity. The amount at the base date is obtained through multiplication of the base price by the current quantity.

$$\text{IPaasche} = \frac{\sum_{i=1}^{n} P_t^i * Q_t^i}{\sum_{i=1}^{n} P_0^i * Q_t^i} \tag{4.2}$$

where:

IPaasche = Paasche Index;
P_t = price of the good(s) in period t (current);
Q_t = quantity of the good(s) in period t (current);
P_0 = price of the good(s) at the base date.

4.3 FISHER INDEX (1922)

The Fisher Index (1922) is an alternative indicator of the general growth, decline or no alteration in the quantities of inputs used or products manufactured. In its original formulation, this index is the geometric mean of the Laspeyres and Paasche Indexes of quantity. The Fisher Index is founded on the divergence between the Laspeyres and Paasche Index values, the first being an overestimate, while the second underestimates the real value (Dumagan, 2002). The Fisher Index (1922) is shown in Equation 4.3.

$$\text{IFisher} = \sqrt{\text{ILaspeyres} * \text{IPaasche}} \tag{4.3}$$

where:

IFisher = Fisher Index;
ILaspeyres = Laspeyres Index;
IPaasche = Paasche Index.

4.4 DROBISH INDEX

The Drobish Index is the simple arithmetic mean between the Laspeyres and Paasche Indexes, as shown in Equation 4.4.

$$\text{IDrobish} = \frac{\text{ILaspeyres} + \text{IPaasche}}{2} \tag{4.4}$$

where:
    IDrobish = Drobish Index;
    ILaspeyres = Laspeyres Index;
    IPaasche = Paasche Index.

4.5 TÖRNQVIST INDEX (1936)

The Törnqvist Index (1936) may be obtained by the weighted means of geometric growth rates of data on quantity and relative prices. The Törnqvist Index (1936) uses the prices as much for the base date as for the period of comparison. Thus, the prices vary from year to year over the whole period analyzed, and this can, in certain cases, be considered to be a disadvantage due to the non-availability of data on product prices for all the years analyzed (Kohli, 2004).

The Törnqvist Index (1936) deals with variations, considering the Total Productivity of the Factors (TPF), which is the difference between the production growth rate and the growth rate of the inputs. In other words, the alterations in quantities of the product, which may not be explained by the variations in the use of the inputs, are the productivity gains. The Törnqvist Index (1936) (Equation 4.5) has been used in several areas, such as agriculture, industry and infrastructure (Kohli, 2004).

$$\ln(PTF_{pt}/PTF_{pt-1}) = \frac{1}{2}\sum_{i=1}^{n}(S_{it} + S_{it-1})\ln\left(\frac{Y_{it}}{Y_{it-1}}\right)$$
$$-\frac{1}{2}\sum_{j=1}^{m}(C_{jt} + C_{jt-1})\ln\left(\frac{X_{jt}}{X_{jt-1}}\right) \tag{4.5}$$

where:
    $\ln(PTF_{pt}/PTF_{pt-1})$ = the variation in the Total Productivity of the Factors between
        period t and period t−1;
    Y_{jt} = quantity of products;
    X_{jt} = quantity of inputs;
    S_{it} = prices of the products;
    C_{jt} = prices of the inputs.

Thus, Equation 4.5 is divided into two distinct indexes. The first is the aggregate index of the products, weighted according to the participation of each product in the total value of the production. The second is the aggregate index of the inputs, weighted by the participation of each input in the total cost. For example, when the index of the inputs grows more than that of the products, there is a fall in the TPF. When the index of the products grows more than the inputs, there is an increase in the TPF. However, the Törnqvist Index (1936) uses the prices as a weighting factor, and this may lead to limitations for the analysis, mainly in analyses of periods that are longer or with high inflation.

4.6 FREE DISPOSAL HULL (FDH)

The Free Disposal Hull (FDH) model is an alternative technique for DEA (the concepts of DEA will be discussed in depth later). FDH was introduced and developed by Deprins et al. (1984). FDH is based on a representation of production technology provided by the production plans observed, imposing a high degree of availability of inputs and outputs, but without the hypothesis of convexity (Leleu, 2006). In the convexity hypothesis, the means of the sample analyzed are preferable, in detriment to the extreme values.

For example, with a certain budget, a buyer of a company has the possibility of acquiring two distinct raw materials (A and B). In this case, he/she may opt to acquire 100 units of raw material A and no raw material B. He may also acquire 100 units of B and no A. Another option is acquisition of 50 units of each. According to the presupposition of convexity, the buyer would opt for the latter option (50 units of each raw material, A and B), as needs would better be met with a little of each raw material, than with a lot of one and none of the other.

FDH is used to establish a group of best practices among a set of observed units and to identify the units that are inefficient when compared to the group of best practices. Thus, a comparison of efficiency may be made only on the basis of the performance actually observed. Units of hypothetical best practices are not permitted (Benslimane & Yang, 2007). If FDH were compared to DEA, the fundamental difference between these techniques concerns the supposition of convexity. Whereas DEA imposes convexity on the production set, the FDH estimator does not exert any such restriction on this aspect. However, the supposition (or not) of convexity may lead to totally different results, regarding the efficiency of the units analyzed. Thus, the use of DEA with the supposition of convexity is preferable in relation to FDH.

4.7 THICK FRONTIER APPROACH (TFA)

The TFA (Thick Frontier Approach) is a parametric technique that assumes that deviations from the predicted performance values represent the inefficiency. Thus, the analyses utilizing TFA assume that the companies with best practices do not have levels of efficiency that deviate significantly from the frontier, so that their efficiency is constant throughout the period analyzed. The central assumption is that there is a group of companies that are efficient in the period analyzed (Bauer et al., 1998).

TFA assumes that the performance value deviations predicted in the quartiles of higher and lower performance of the companies represent only a random error. Besides this, it considers that the deviations in the predicted performance between the quartiles of highest and lowest mean costs represent only inefficiency. In the majority of the applications, TFA provides an estimate of the differences in efficiency between the highest and the lowest quartile to indicate the general level of efficiency but does not provide accurate estimates of efficiency for each of the individual companies (Bauer et al., 1998).

In order to illustrate this limitation of TFA in not providing accurate estimates of efficiency for each company, we may cite an analysis of efficiency in the banking sector, considering 10 distinct bank branches over time. In this case, TFA will provide the efficiency indexes of only the best and worst branches. If it is necessary to obtain an estimate of efficiency for each branch in each period (each month, for example), TFA is not recommended.

4.8 DISTRIBUTION-FREE APPROACH (DFA)

The Distribution-Free Approach (DFA) is a parametric technique that specifies a functional form for the efficient frontier, as well as other techniques, such as TFA. However, DFA separates the inefficiency of random errors differently from TFA. In this sense, DFA assumes that there exists central efficiency or mean efficiency for each company that is constant over time, whereas random error tends to increase the mean over time (Berger, 1993). Contrary to other approaches, a set of panel data is necessary.

Just as happens with the other approaches to efficiency, there is concern that the levels of the efficiency estimates of DFA may be influenced by arbitrary presuppositions. The measurement of the efficiency of the nucleus means that the efficiency variations over time for an individual company tend be calculated as a mean with random errors. DFA also implicitly assumes that inefficiency is the only fixed effect that does not vary over time (Bauer et al., 1998). However, if there are other non-fixed factors that can influence a company's costs, such as an unplanned halt for maintenance for a certain period of time, this effect is not considered in the inefficiency.

4.9 STOCHASTIC FRONTIER ANALYSIS (SFA)

Stochastic Frontier Analysis (SFA) takes a parametric approach and is directly linked to the econometric theories. The technique was initially proposed by Aigner et al. (1977) and is based on efficiency studies developed by Farrell (1957). Econometrics, which may literally be expressed as "economic measurement", is related to the use of mathematics and statistics in the economic theories, with the aim of giving empirical support to the economic models (Gujarati & Porter, 2011).

The principal objective of SFA is to estimate a function to obtain maximum production by combining the factors efficiently. This technique involves a specific function of production for data that are analyzed transversally (cross-sectionally), with an error term having two components, one being to account for the random effects and the other to account for the technological inefficiency.

The stochastic concept, originating from the term, stochastic frontier, may be defined as a study aimed at application of the calculation of probabilities to statistical data, in such a way as to establish the existence of permanent regular variables. Thus, SFA assumes a stochastic relation between inputs and outputs. Besides this, SFA provides support that not all the existing deviations in relation to the frontier represent inefficiency, but may also be noise in the data (Bogetoft & Otto, 2010). The stochastic frontier models manage to recognize that the existing deviations in relation to the frontier may represent two causes: external factors without control or arising from technical inefficiency (Bogetoft & Otto, 2010).

According to Folland et al. (2009), the approach proposed by the SFA model treats each DMU in a unique manner and arises from the presupposition that it may be affected by an unexpected fact. For example, a hospital, operating at optimum levels of efficiency for a period, lost its main supplier, taking some months to acquire another with the same charges, reliability and rapidity, among other parameters. In this case, despite the company presenting an optimal management model, its efficiency was affected by external factors. Thus, if a DMU were affected by some unexpected fact during the period of analysis, its production efficiency and costs would be affected, randomly displacing its efficiency frontier (Folland et al., 2009). Finally, it is to be stressed that the SFA technique is commonly used in econometric analyses and little used in other contexts, such as for the analysis of productivity and efficiency in production systems.

4.10 ANALYTIC HIERARCHY PROCESS (AHP)

AHP is a technique, developed by Saaty (1977), used to solve complex problems. The technique is applied to the solution of unstructured problems in the scientific, economic, social and managerial areas, among which there are productivity and efficiency calculations (Mahalik & Patel, 2010). Basically, AHP compares independent criteria and weighs alternatives in a paired mode. Thus, AHP is a weighting technique whereby subjective assessments are obtained with numerical values on a scale. The value of each element obtained forms a judgment matrix, which reflects the relative weight between two comparable factors (Mathivathanan et al., 2017).

In applying the AHP technique, it is necessary to disaggregate the complex problem into a hierarchical structure of characteristics and multi-level criteria, having decision alternatives at the lowest level (Saaty, 1977). For application of AHP, the following steps are suggested (Thanki et al., 2016):

i) Construction of the decision hierarchy.
ii) Comparison among the hierarchical elements.
iii) Analysis of the priority relative to each criterion (normalizing the values of the comparison matrix and obtaining the priorities vector).
iv) Assessment of the consistencies of the relative priorities.
v) Construction of the paired matrix for each criterion, considering each alternative selected.
vi) Obtaining the composite priority for the alternatives.
vii) Choice of the alternative.

4.10.1 CONSTRUCTION OF THE DECISION HIERARCHY

Initially, the context to be analyzed is structured into hierarchical levels to facilitate its assessment and understanding. Thus, it is necessary, after definition of the study objective, for the criteria and defined alternatives to be structured hierarchically, as in Figure 4.1. This structuring aids the overall visualization of the study context.

4.10.2 COMPARISON AMONG THE HIERARCHICAL ELEMENTS

Afterward, it is necessary to establish priorities among the elements for each level of the hierarchy, by means of a comparison matrix. Thus, the first factor to consider is the determination of a scale of values for comparison, which should not exceed a total of nine points, as presented in Chart 4.2.

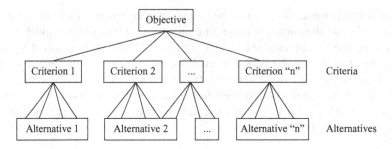

FIGURE 4.1   Basic hierarchical structure of AHP.

CHART 4.2
Saaty Scale 1–9 for Comparison of Pairs

Intensity of the weight	Definition	Explanation
1	Equally important	Two activities contribute equally to the objective
3	Moderately important	Experience and judgment slightly favor one over the other
5	Very important	Experience and judgment greatly favor one over the other
7	Intensely important	One activity is strongly favored, and its domination is demonstrated in practice
9	Extremely important	The importance of one over the other is affirmed in the highest possible order
2,4,6,8	Intermediate weights	Used to represent the commitment among the priorities listed above

---

When two elements, for example, A and B, are compared by the decision-maker, the value 1 signifies that there is no difference between A and B. However, if A is extremely important/preferable in relation to B, the value 9 is used for A. If the opposite is true, the same criteria are used. One of the recurrent criticisms of AHP is that the criteria stipulated are based on the decision-makers' subjective perceptions. This may cause bias in the results of the context analyzed.

After the paired comparisons are made through the previously created hierarchy, a priority matrix should be developed (Figure 4.2), on which the results of comparisons among the elements analyzed are presented. In the matrix shown in Figure 4.2, the elements obey the following rule: $a_{ij} = 1/a_{ji}$, where a_{ij} are real positive values, i is the index that represents the row and j the column. These characteristics make the matrix designated as the positive reciprocal. It can be perceived that the diagonal of the matrix is all unitary, after all each attribute, compared to itself, is equal to the unit.

Based on the weights supplied by the assessor, the matrix was established, as represented in Figure 4.3, where:

W_m = Weight referring to the row.
W_n = Weight referring to the column.

$$A = \begin{bmatrix} 1 & a_{12} & \cdots & a_{1n} \\ 1/a_{21} & 1 & \cdots & a_{2n} \\ \vdots & \vdots & \vdots & \vdots \\ 1/a_{n1} & 1/a_{n2} & \cdots & 1 \end{bmatrix}, \text{where}$$

$$a_{ij} > 0 \Rightarrow positive$$
$$a_{ij} = 1 \qquad a_{ij} = 1$$
$$a_{ij} = 1/a_{ij} \Rightarrow reciprocal$$
$$a_{ik} = a_{ij} \cdot a_{jk} \Rightarrow consistency$$

FIGURE 4.2   General matrix of AHP priority.

$$\begin{array}{cccc} & F_1 & F_2 & \cdots & F_n \end{array}$$

$$\begin{array}{c} F_1 \\ F_2 \\ \vdots \\ F_n \end{array} \begin{bmatrix} w_1/w_1 & w_1/w_2 & \cdots & w_1/w_n \\ \dfrac{1}{w_2/w_1} & w_2/w_2 & \cdots & w_2/w_n \\ \vdots & \vdots & \vdots & \vdots \\ \dfrac{1}{w_m/w_1} & \dfrac{1}{w_m/w_2} & \cdots & w_m/w_n \end{bmatrix}$$

FIGURE 4.3   General matrix of AHP priority with weights for the criteria.

In order to facilitate comprehension of Figures 4.2 and 4.3, we show their evolution step by step, by considering the example of Arueira (2014) with five criteria (A to E) and seven alternatives (1 to 7). Chart 4.3 shows an example of a priority matrix, in which the five criteria are compared in pairs.

In the comparison of the pairs AA, BB, CC, DD and EE, the value will always be equal to 1, because, when compared with itself, the criterion possesses the same importance. For the other comparisons, the scale presented in Chart 4.2 is used. Another aspect to be observed is that the result of the comparison of item A with item C is the inverse of the comparison of item C in relation to item A; if A has a

CHART 4.3

Matrix AHP Priority With Five Criteria

Criteria	A	B	C	D	E
A	1	3	1/9	1/5	5
B	1/3	1	1/9	1/7	1
C	9	9	1	3	7
D	5	1/3	1/3	1	9
E	1/5	1	1/7	1/9	1

value equal to 1/9, when compared with C, C should have a value equal to 9/1 when compared with A. This occurs because the method uses the concept of reciprocity, where the comparison between the pairs is realized only once. This concept is of greater use in implementing AHP, reducing the chance of inconsistencies occurring (Arueira, 2014).

4.10.3 ANALYSIS OF THE RELATIVE PRIORITY OF EACH CRITERION

In order to obtain the relative priority of each criterion, it is necessary to: (i) normalize the values of the matrix of comparisons, the objective being to make all the criteria equal to the same unit; and (ii) obtain the priorities vector, the objective being to identify the order of importance of each criterion. Normalization of the data of the priority matrix is obtained by the ratio of the value of each component to the sum of all the components of the columns, as shown in Chart 4.4.

With normalization of the data, it becomes possible to calculate the mean of the normalized values so that the priority vector of each criterion is defined (Chart 4.5).

It is clear that criterion C is the one that possesses the largest priority vector (0.55). Thus, its result should exert greater impact on the central objective of the analysis.

CHART 4.4
Normalization of the Priority Matrix

Criteria	A	B	C	D	E
A	0.0644	0.2093	0.0654	0.0449	0.2174
B	0.0215	0.0698	0.0654	0.0321	0.0435
C	0.5794	0.6279	0.5888	0.6736	0.3043
D	0.3219	0.0233	0.1963	0.2245	0.3913
E	0.0129	0.0698	0.0841	0.0249	0.0435
Totals	1.00	1.00	1.00	1.00	1.00

CHART 4.5
Priority Vector

Criteria	A	B	C	D	E	Priority Vector
A	0.0644	0.2093	0.0654	0.0449	0.2174	**0.12**
B	0.0215	0.0698	0.0654	0.0321	0.0435	**0.05**
C	0.5794	0.6279	0.5888	0.6736	0.3043	**0.55**
D	0.3219	0.0233	0.1963	0.2245	0.3913	**0.23**
E	0.0129	0.0698	0.0841	0.0249	0.0435	**0.05**
Totals	1.00	1.00	1.00	1.00	1.00	1.00

4.10.4   ASSESSMENT OF THE CONSISTENCY OF THE RELATIVE PRIORITIES

In the assessment of the consistency of the relative priorities, the sum should be made of the multiplication of the values stipulated for the criteria in the priority matrix (Chart 4.3) by the priority vector of each criterion. In addition, a calculation should be made of the ratio of the results obtained with the sum, to the vector of each criterion, as shown in Equation 4.6.

$$AW = \lambda_{max} W \tag{4.6}$$

where:

A = comparison made in the priority matrix;
W = priority vector.

In the example illustrated, the following values are obtained:

$$A = (0.12 * 1 + 0.05 * 3 + 0.55 * 0.11 + 0.23 * 0.20 + 0.05 * 5)/0.12 = \textbf{5.22}$$

$$B = \big(0.12 * 0.33 + 0.05 * 1 + 0.55 * 0.11 + 0.23 * 0.14 + 0.05 * 1\big)/0.05 = \textbf{4.65}$$

$$C = \big(0.12 * 9 + 0.05 * 9 + 0.55 * 1 + 0.23 * 3 + 0.05 * 7\big)/0.55 = \textbf{5.67}$$

$$D = \big(0.12 * 5 + 0.05 * 0.33 + 0.55 * 0.33 + 0.23 * 1 + 0.05 * 9\big)/0.23 = \textbf{6.43}$$

$$E = \big(0.12 * 0.2 + 0.05 * 1 + 0.55 * 0.14 + 0.23 * 0.11 + 0.05 * 1\big)/0.05 = \textbf{4.53}$$

Afterwards, the ratio between the sum of the values calculated in this new vector and the number of criteria should be calculated. With this, an approximation of the result of the maximum autovalue (λ_{max}) is obtained. Thus, the closer the maximum autovalue is to the number of criteria, the more consistent is the result obtained. The autovalue (λ_{max}) is obtained by means of Equation (4.7):

$$\lambda_{max} = \frac{1}{n} \sum_{i=1}^{n} \frac{[Aw]}{wi} \tag{4.7}$$

where:

λ_{max} = maximum autovalue;
Aw = resulting matrix of the product of paired comparison by (wi), the matrix of weights to be processed;

In the example illustrated, the following values are obtained:

$$\lambda_{max} = (5.22 + 4.65 + 5.67 + 6.43 + 4.53)/5 = 5.30$$

Thus, the index consistency utilizing Equation (4.8) is calculated

$$IC = \frac{\lambda_{max} - n}{(n-1)} \tag{4.8}$$

where:

IC = consistency index;

λ_{max} = maximum autovalue;

n = number of criteria.

In the example illustrated, the following values are obtained:

$$IC = (5.30 - 5)/4 = 0.075$$

With the definition of IC, it is possible to verify the standard random index proposed by Saaty (1991) in Chart 4.6, whereas n denotes the number of criteria analyzed, and IR, the random index. In the example illustrated, the number of criteria assessed (n) is equal to five. Thus, IR = 1.12.

Finally, the consistency ratio (RC) (Equation 4.9) is calculated, which is the ratio between the consistency index (IC) and the random index (IR). The matrixes that possess a consistency ratio (RC) below 0.10 are considered to be acceptable. In the example used, the matrix proposed may be accepted, as the result of RC = 0.067

$$RC = \frac{IC}{IR} \tag{4.9}$$

where:

RC = consistency ratio;

IC = consistency index;

IR = random index.

In the example illustrated, the following values are obtained:

$$RC = (0.075/1.12) = 0.067$$

4.10.5   CONSTRUCTION OF THE PAIRED MATRIX FOR EACH CRITERION, CONSIDERING EACH OF THE ALTERNATIVES SELECTED

After assessment of the consistency of the relative priorities, all the procedures for construction of the comparison matrix and for determination of the relative priority of each criterion should be made again. In this step, the relative importance of each alternative comprising the hierarchical structure of the problem under analysis should be observed.

CHART 4.6

Standard Chart of IR Calculation

N	1	2	3	4	5	6	7	8	9	10	11	12	13	14	15
IR	0.00	0.00	0.58	0.90	1.12	1.24	1.32	1.41	1.45	1.49	1.51	1.48	1.56	1.57	1.59

Source: Saaty (1991).

4.10.6   OBTAINING THE COMPOSITE PRIORITY FOR THE ALTERNATIVES

After realization of this procedure, there should be analysis of the general score of the composite priorities of the alternatives, obtained by multiplication of the priority vectors obtained in the previous steps by the values of the relative priorities, as obtained at the start of the method. Chart 4.7 lists the values of the example presented with final calculations, without describing every step of the calculation of each criterion so as not to be exhaustive and repetitive.

In order to continue with the process, it is necessary to normalize the values found, putting all of them into the same base (from zero to 1) to enable calculation of the final score of each alternative. Thus, calculation is made of the ratio of the value observed of each alternative to the total value of all the alternatives, when the indicator is of the type where the bigger, the better. However, when the indicator is of the type where the smaller, the better, it will be necessary to perform the normalization in two phases: (i) calculation of the ratio of the sum of the values of all the alternatives for that criterion to the observed value of each alternative; and (ii) with the new values obtained, calculation of the ratio of each alternative to the total value of all the alternatives for that defined criterion. The results of the calculations described in the example illustrated are presented in Chart 4.8.

4.10.7   CHOICE OF THE ALTERNATIVE

After normalization of the data, it is possible to calculate the score of each alternative, utilizing the priority vectors found in the previous steps. For this, it is necessary to multiply the normalized values by the vector found for each criterion. In the example used to illustrate the functioning of AHP, the best alternative presented is number two, with a total score of 0.1748 (Chart 4.9).

The AHP method may be used, for example, to carry out an assessment of performance for selection of suppliers. In this case, based on particular criteria (for example, performance in terms of delivery, cost and quality), the supplier that obtains the

CHART 4.7
Values Observed

| Alternatives | Criteria | | | | |
	A	B	C	D	E
1	10.00	1.00	0.70	0.68	4.00
2	6.00	0.80	0.53	0.54	4.00
3	7.00	0.90	0.81	0.60	3.00
4	10.00	0.95	0.93	0.48	3.00
5	5.00	0.70	0.62	0.57	4.00
6	9.00	0.85	0.94	0.50	5.00
7	9.00	0.90	0.62	0.61	3.00
TOTAL	56.00	6.10	5.15	3.98	26.00

CHART 4.8
Normalized Values

Alternatives	Criteria				
	A	B	C	D	E
1	0.11	0.16	0.14	0.12	0.15
2	0.18	0.13	0.19	0.15	0.15
3	0.15	0.15	0.12	0.13	0.12
4	0.11	0.16	0.11	0.17	0.12
5	0.21	0.11	0.16	0.14	0.15
6	0.12	0.14	0.12	0.16	0.19
7	0.12	0.15	0.16	0.13	0.12
TOTAL	1.00	1.00	1.00	1.00	1.00

CHART 4.9
Ranking of the Alternatives

Alternative	Score
2	0.1748
5	0.1586
7	0.1460
6	0.1335
1	0.1331
3	0.1273
4	0.1267

best score may be considered the most efficient and may be the one chosen for supply in a certain negotiation.

However, despite being used for analysis of productivity and efficiency, the focus of the analyses utilizing AHP is the hierarchical structuring of complex problems. For analyses of productivity and efficiency, AHP presents limitations, such as the non-provision of targets and slack to establish improvement goals. Besides this, AHP is used for one-off analysis, although it is not the most recommended technique for analysis of the productivity and efficiency of a production system which possesses defined inputs and outputs.

4.11 OVERALL EQUIPMENT EFFECTIVENESS (OEE)

OEE is a metric used to measure the individual performance of the equipment in industrial companies. OEE was developed based on Total Production Maintenance (TPM) concepts introduced by Nakajima (1989). Its role is to aid identification and measurement of losses in operations. The fundamental idea of OEE is based on the conception that the ideal potential of the equipment is reduced by losses that occur in its operation. Chart 4.10 describes the losses measured by OEE.

To define the OEE value, the relation between the time taken to add value and the total time available for production of a certain resource is calculated. If it is considered that the resource being analyzed is causing a bottleneck (that is, its capacity is less than its expected demand), the index, Total Effective Equipment Productivity (TEEP), is used. This index considers that the resource would not have scheduled interruptions, and, therefore, all the stoppages are relevant to the performance calculation. If the equipment does not cause a bottleneck, the OEE Index is considered, the scheduled halts being subtracted from the calculation denominator.

OEE is the result of multiplication of three indexes. The μ_1 or Index of Operational Time Availability (ITO), represents the percentage of time (initially programed for the operation) in which the equipment effectively worked, subtracting any stoppages. The μ_2 or Index of Operational Performance (IPO) corresponds to the percentage of exploitation of the time in which the machine effectively operated for production, taking the theoretical cycle as the base time.

Finally, μ_3, or Index Approved Products - Quality (IAP) corresponds to the percentage of production time in which the equipment effectively produced good items.

So as to clarify the TEEP and OEE concepts, Figures 4.4 and 4.5 illustrate the relationships of the times used and controlled for proper use of each index.

CHART 4.10
Losses Measured by OEE

Type of loss	Description of Loss
Equipment breakdowns	Unscheduled equipment stoppages. Generally, occur due to mechanical problems that require intervention to re-establish their operation.
Setup and regulation	Arise from production programing and product mix transformed by the equipment. In this case, the variation of the mix is high and the production batches small, and the losses per setup tend to reduce the productive capacity of the equipment.
Short halts and idle periods	Minor equipment stoppages that can be dealt with easily, allowing prompt resumption of production. Losses due to idle periods caused by shortage or interruption of material supplies.
Speed reduction	Characterized by decrease in the pace of production in relation to the goal. Various possible causes: equipment failure, operators' lack of experience, raw material of low quality or not conforming to specifications, among others.
Lack of quality and rework	In these cases, the equipment produces at the ideal speed, but the products do not match the quality standard. These losses may be transformed into waste or require corrective rework.
Performance decline	Generally, occur in processes for resumption of production, such as when new product launches, return of laid-off employees, at the start of a shift or at the beginning of the working week.

TEEP – Total Effective Equipment Productivity

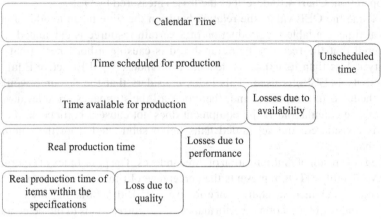

FIGURE 4.4   Illustration of TEEP.

OEE – Overall Equipment Effectiveness

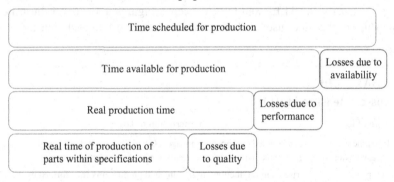

FIGURE 4.5   Illustration of OEE.

To calculate the OEE, it is necessary to time the product cycle in the operation analyzed, stipulate a routine for logging the operation, and define the stoppage typology. Proper application of OEE may be achieved through: i) constant updating of the product cycle times; ii) constant updating of the equipment log; and iii) daily analysis of the calculation of OEE and its indexes. Chart 4.11 shows the OEE concepts, meanings and equations.

At its origin, OEE represented an advance in the way industrial companies used to measure the performance of their equipment. This was because, until the development of OEE, only the availability was considered in the use of the equipment, which resulted in oversizing the productive capacity. However, the objective of OEE is linked directly to the measurement of the equipment individually, ignoring its relations with other equipment or even with the persons involved in its operation (Muchiri & Pintelon, 2008). For these motives, decisions cannot be taken that involve other elements, besides the equipment itself, based only on the OEE result

CHART 4.11
Synthesis of the OEE Concepts

Variable	Meaning	How to Calculate	Example
Q_i	Number of items produced within the specifications	Record the quantity of parts of a particular item i produced within the specifications	50 parts of item A and 20 parts of Item B
QN_i	Number of items produced outside the specifications (rejects)	Record the quantity of parts of a particular item i processed outside the specifications	10 parts of item A 0 parts of item B
Tp_i	Cycle time	Measure in minutes the time necessary to produce a part of item i	Item A – 2 minutes/part Item B – 5 minutes/part
$Tp_i * Q_i$	Ideal added value time.	Multiply the cycle time of the part by the number of parts produced within the specifications <center>$Tp_i * Q_i$</center>	For item A: 50 * 2 = 100 minutes For item B: 20 * 5 = 100 minutes
$Tp_i * QN_i$	Theoretical time lost with items outside the specifications.	Multiply the cycle time of the part by the number of items produced outside the specifications <center>$Tp_i * QN_i$</center>	For item A: 10 * 2 = 20 minutes For item B: 0 * 10 = 0 minutes
PP	Scheduled halt time	Time of a scheduled halt in minutes	Halt for cleaning: 10 minutes Halt for a daily meeting: 15 minutes
PN	Unscheduled halt time	Log the time of an unscheduled halt in minutes	25 minutes for setup and 20 minutes for corrective maintenance
$\sum TC$	Calendar time	Define the total time of the daily availability of production of the resource. This information should be defined daily.	480 minutes
$\sum Tp_i * Q_i$	Sum of the ideal time for added value.	Consider the total time used for production of items within the specifications	100 minutes of value added to part A 100 minutes of value added to part B = 200 minutes
$\sum Tp_i * QN_i$	Sum of the theoretical time lost with items outside the specifications.	Consider the total time used for production of items outside the specifications	20 + 0 = 20 minutes
$\sum PP$	Sum of the scheduled stoppage times	Sum of the scheduled halt times <center>$\sum PP$</center>	10 + 15 = 25 minutes

(Continued)

CHART 4.11 (CONTINUED)
Synthesis of the OEE Concepts

Variable	Meaning	How to Calculate	Example
$\sum PN$	Sum of the unscheduled stoppage times	Sum of the unscheduled halt times $$\sum PN$$	$25 + 20 = 45$ minutes
TEEP OEE	If resource causes a bottleneck (demand > capacity), use TEEP. If the opposite, use OEE.	TEEP – Use calendar time as the base, without subtracting scheduled halts: $$TEEP = \frac{\sum Tp_i * Q_i}{\sum TC}$$ OEE – Use calendar time as the base, subtracting scheduled halts: $$OEE = \frac{\sum Tp_i * Q_i}{\sum TC - \sum PP}$$ OEE is also equal to a: $$\mu_1 * \mu_2 * \mu_3$$	$TEEP = \dfrac{200}{480} = 41.67\%$ $OEE = \dfrac{200}{480 - 25} = 43.96\%$
μ_1 (Availability	Availability Index. Represents the percentage of activation time of the resource, in comparison to resource time available.	If the resource causes a bottleneck: $$\frac{\sum TC - \sum PP - \sum PN}{\sum TC}$$ If the resource causes a bottleneck: $$\frac{\sum TC - \sum PP - \sum PN}{\sum TC - \sum PP}$$ or simply $$\frac{\text{Activation time of the resource}}{\text{Total time available for the operation}}$$	If the resource causes a bottleneck: $\mu_1 = \dfrac{480 - 25 - 45}{480}$ $\mu_1 = \dfrac{410}{480} = 85.42\%$ If the resource does not cause a bottleneck: $\mu_1 = \dfrac{480 - 25 - 45}{480 - 25}$ $\mu_1 = \dfrac{410}{455} = 90.11\%$
μ_2 (Performance)	Performance Index. Represents the percentage of activation time of the resource that effectively added value to the part, considering its cycle time as a parameter.	$$\frac{\sum Tp_i * Q_i + \sum Tp_i * QN_i}{\sum TC - \sum PP - \sum PN}$$ or simply $$\frac{\text{Theoretical production time}}{\text{Operation time of the machine}}$$	$\mu_2 = \dfrac{200 + 20}{480 - 25 - 45}$ $\mu_2 = \dfrac{220}{410} = 53.66\%$

(Continued)

CHART 4.11 (CONTINUED)
Synthesis of the OEE Concepts

Variable	Meaning	How to Calculate	Example
μ_3 (Quality)	Quality Index. Represents the percentage of time spent on production of items within the specifications in relation to the total production time.	$$\dfrac{\sum Tp_i * Q_i}{\sum Tp_i * Q_i + \sum Tp_i * QN_i}$$ or simply $$\dfrac{\text{Theoretical production time for items within the specifications}}{\text{Theoretical production time}}$$	$\mu_3 = \dfrac{200}{200+20}$ $\mu_3 = \dfrac{200}{220} = 90.91\%$

(Andersson & Bellgran, 2015). This means that, even being a broadly divulgated metric in manufacturing companies, its application does not completely reflect the needs of all companies, in that, by using OEE, productivity and efficiency cannot be analyzed from a systemic perspective.

Besides this, OEE does not consider the reductions in the scheduling that would make the equipment be used less, and it also does not consider the losses related to the use of materials. The OEE calculation formula does not consider, for example, the expenses necessary to obtain the calculated indexes, nor the proportion of inputs consumed to obtain these production levels (Jonsson & Lesshammar, 1999). OEE assesses the production operations through the results obtained with the use of the equipment, without considering the other components of these operations. For example, a certain production schedule may improve the OEE of the equipment but entail a greater quantity of rejects or leftover raw material.

Only the measuring of OEE may not be considered sufficient to establish goals for improvement programs. If considered in this way, the decisions taken may favor only optimal places, which do not necessarily reflect improvement for the entire company or the production process. Thus, it is considered that other aspects not considered by OEE exist that are part of the context of manufacturing and services, such as energy efficiency, raw material efficiency, production costs and delivery priorities (Jeong & Phillips, 2001; Oechsner et al., 2002).

4.11.1 OTHER METRICS FOR ASSESSMENT OF EQUIPMENT EFFICIENCY

Due to the limitations described, there are some proposals for indicators that seek to complement or substitute for OEE. Of the performance measurement metrics presented in Chart 4.12, only PEE, EPR and CUB are comparable to OEE in relation to the production unit measured, that is, both are used to measure the individual performance of equipment. The other tools make measurements of production cells or the entire fabrication process, and, therefore, present distinct OEE characteristics.

The techniques shown in Chart 4.12 contribute to the improvement of the assessment systems of the performance of operations in production systems. However, as

CHART 4.12

Alternative Indicators Proposed for Complementation of OEE

Indicator	Description of indicator	Unit of analysis	Authors
OFE - Overall Factory Effectiveness	Measures combined activities where there are relations among different machines and equipment.	Production Cells	Oechsner et al. (2002)
OTE - Overall Throughput Effectiveness	Used to measure factory performance and identify bottlenecks and hidden capacities.	Factory	Muthiah and Huang (2007)
GPE - Overall Production Effectiveness	Uses the sequence of individual measurements combined to determine the performance of the system after each integration among the processes.	Factory	Lanza et al. (2013)
PEE – Production Equipment Effectiveness	Uses the same OEE indicators, although it attributes different weights to each, according to their importance in the process being measured.	Factory	Raouf (1994)
OPE – Overall Plant Effectiveness	Proposes measuring the real outputs of the factory in relation to the predicted outputs.	Factory	Scott and Pisa (1998)
OAE – Overall Asset Effectiveness	Proposes measurement of the factory's real outputs in relation to the forecast outputs.	Factory	Neely et al. (1995)
EPR - Equipment Performance Reliability	Measures the reliability of the equipment, related to its capacity to fulfil the technical characteristics for which it was designed.	Equipment	Muchiri and Pintelon (2008)
CUB - Capacity Utilization Bottleneck	Measures the outputs in a bottleneck situation in relation to the theoretical production that should occur.	Equipment	Konopka (1995)

in the example of OEE, these also do not consider important elements for assessment of the operational performance, such as raw materials, maintenance expenses or stock of spare parts, among others.

Improvement of the operational performance has been, since the 1990s, one of the central issues to allow companies to remain competitive (Bititci et al., 2012). For this reason, the problems involving definition of the correct metrics for assessment of the operational performance continue to be the object of studies in the academic and business areas (Bennett et al., 2014).

4.12   INTRODUCTION TO DATA ENVELOPMENT ANALYSIS (DEA)

Among the techniques for analysis of productivity and efficiency, DEA may be a highlight. The history of the development of DEA began with Edward Rhodes' doctoral thesis in 1978, under the supervision of W.W. Cooper. The objective of the research was to assess the results of an assistance program for needy students, based in American public schools, with federal government support. The central idea was to compare the performance of a set of students from schools that participated in the program, with another set of students from non-participant schools. The performance of the students was measured in terms of defined products, for example, increase in self-esteem of the needy children, and inputs, such as time spent by the mother on reading exercises with her children (Charnes et al., 1978).

This attempt to estimate the technical efficiency of schools, based on multiple inputs and products, resulted in the formulation of the CCR (an acronym based on Charnes, Cooper and Rhodes, the surnames of the authors) model of Data Envelopment Analysis. This work was first published in 1978.

In analyses using DEA, an efficiency (or maximum productivity) curve is obtained considering the optimal relation between inputs and products. This curve can be defined as an efficiency frontier. Thus, the units considered to be efficient will lie on this curve, while the inefficient will be located below it. The frontier will provide the parameters for an inefficient unit of analysis to become efficient.

As with linear programming, using DEA it is possible to determine the maximum output quantities produced for a certain level of input consumption. In addition, with DEA, it is possible to determine which DMUs have reached these maximum levels (DMU concepts will be detailed later), these being positioned at the efficiency frontier, and also which input or output parameters would need to be improved in the other DMUs so that they can reach this efficiency frontier. DMUs positioned at the efficiency frontier are considered benchmarking, and the adjustments required for those DMUs that were positioned outside the efficiency frontier can be detailed, based on the calculation of the targets and slack. This book proposes DEA as the main technique to be used for analysis and management of productivity and efficiency in production systems. There are attributes that make DEA quite operational, such as the following: the relations between multiple inputs and outputs can be transformed into a single efficiency index, and there is a possibility of identifying input savings or production increases for the inefficient DMUs to be directed toward the efficient ones.

The main advantage of DEA is that the technique does not need any restriction on the functional form of the production relationship between inputs and outputs. In this sense, the inputs and outputs to be used in the DEA model do not need, for example, to obey a standard of equal measurement units. This characteristic makes DEA very flexible and adaptable for several different contexts, to be considered whether it be for analysis of production systems or economic analysis.

DEA also does not require any supposition for the distribution of the term "inefficiency", and, as a result, can be considered as a technique of a deterministic nature. In other words, all deviations from the efficiency frontier are considered to be under the company's control, and therefore attributable to the term, inefficiency (Çelen, 2013). In addition, DEA does not use statistical inferences, or measurements of central tendency, tests of coefficients or formalizations of regression analyses.

According to Hwang et al. (2013), DEA can be used as a decision support technique for performance monitoring in organizations oriented towards productivity and efficiency. Since 1978, studies have been conducted in diverse application areas, using DEA to make performance measurements (Liu et al., 2016). Thus, in both the academic and the business contexts, it is perceived that data envelopment analysis is widely accepted and is being used and developed (Liu et al., 2013).

REFERENCES

Aigner, D., Lovell, C. K., & Schmidt, P. (1977). Formulation and estimation of stochastic frontier production function models. *Journal of Econometrics*, 6(1), 21–37.

Andersson, C., Bellgran, M. (2015). On the complexity of using performance measures: Enhancing sustained production improvement capability by combining OEE and productivity. *Journal of Manufacturing Systems*, 35, 144–154.

Arueira, A. (2014). *Aplicação do Método AHP para Avaliação de Transportadores* (Doctoral dissertation, PUC-Rio).

Balfaqih, H., Nopiah, Z.M., Saibani, N., & Al-Nory, M.T. (2016). Review of supply chain performance measurement systems: 1998–2015. *Computers in Industry*, 82, 135–150.

Bauer, P. W., Berger, A. N., Ferrier, G. D., & Humphrey, D. B. (1998). Consistency conditions for regulatory analysis of financial institutions: A comparison of frontier efficiency methods. *Journal of Economics and Business*, 50(2), 85–114.

Bennett, W., Lance, C. E., & Woehr, D. J. (2014). *Performance measurement: Current perspectives and future challenges*. Psychology Press.

Benslimane, Y., & Yang, Z. (2007). Linking commercial website functions to perceived usefulness: A free disposal hull approach. *Mathematical and Computer Modelling*, 46(9), 1191–1202.

Berger, A. N. (1993). "Distribution-free" estimates of efficiency in the US banking industry and tests of the standard distributional assumptions. *Journal of productivity Analysis*, 4(3), 261–292.

Bititci, U., Garengo, P., Dörfler, V., & Nudurupati, S. (2012). Performance measurement: Challenges for tomorrow. *International Journal of Management Reviews*, 14(3), 305–327.

Bogetoft, P., & Otto, L. (2010). *Benchmarking with Dea, Sfa, and R* (Vol. 157). Springer Science & Business Media.

Çelen, A. (2013). The effect of merger and consolidation activities on the efficiency of electricity distribution regions in Turkey. *Energy Policy*, 59, 674–682.

Charnes, A., Cooper, W. W., & Rhodes, E. (1978). Measuring the efficiency of decision-making units. *European Journal of Operational Research*, 2(6), 429–444.

Coelli, T. J., Rao, D. S. P., O'Donnell, C. J., & Battese, G. E. (2005). *An introduction to efficiency and productivity analysis*. (2nd ed., p. 349). New York: Springer.

Deprins, D., Simar, L., & Tulkens, H. (1984). *Measuring labor efficiency in post offices. The performance of public enterprises concepts and measurements* (pp. 247–263). Amsterdam: Elsevier.

Dumagan, J. C. (2002). Comparing the superlative Törnqvist and Fisher ideal indexes. *Economics Letters*, 76(2), 251–258.

Farrell, M. J. (1957). The measurement of productive efficiency. *Journal of the Royal Statistical Society*, 120(3), 253–290.

Ferreira, C. M. C., & Gomes, A. P. (2009). *Introdução à análise envoltória de dados: teoria, modelos e aplicações*. Viçosa: Editora UFV.

Folland, S., Goodman, A. C., & Stano, M. (2009). *A economia da saúde*. Bookman Editora.

Gujarati, D. N., & Porter, D. C. (2011). *Econometria Básica-5*. AMGH Editora.

Hitt, M.A., Xu, K., & Carnes, C.M. (2016). Resource based theory in operations management research. *Journal of Operations Management*, 41, 77–94.

Hwang, S. N., Chen, C., Chen, Y., Lee, H. S., & Shen, P. D. (2013). Sustainable design performance evaluation with applications in the automobile industry: Focusing on inefficiency by undesirable factors. *Omega*, 41(3), 553–558.

Jeong, K. Y., & Phillips, D. T. (2001). Operational efficiency and effectiveness measurement. *International Journal of Operations & Production Management, 21*(11), 1404–1416.

Jonsson, P., & Lesshammar, M. (1999). Evaluation and improvement of manufacturing performance measurement systems-the role of OEE. *International Journal of Operations & Production Management, 19*(1), 55–78.

Kenyon, G.N., Meixell, M.J., & Westfall, P.H. (2016). Production outsourcing and operational performance: An empirical study using secondary data. *International Journal of Production Economics, 171*, 336–349.

Kohli, U. (2004). An implicit törnqvist index of real GDP. *Journal of Productivity Analysis, 21*(3), 337–353.

Konopka, J. M. (1995). Capacity utilization bottleneck efficiency system-CUBES. *IEEE Transactions on Components, Packaging, and Manufacturing Technology: Part A, 18*(3), 484–491.

Lanza, G., Stoll, J., Stricker, N., Peters, S., & Lorenz, C. (2013). Measuring global production effectiveness. *Procedia CIRP, 7*, 31–36.

Leleu, H. (2006). A linear programming framework for free disposal hull technologies and cost functions: Primal and dual models. *European Journal of Operational Research, 168*(2), 340–344.

Liu, J. S., Lu, L. Y., & Lu, W. M. (2016). Research fronts in data envelopment analysis. *Omega, 58*, 33–45.

Liu, J. S., Lu, L. Y., Lu, W. M., & Lin, B. J. (2013). Data envelopment analysis 1978–2010: A citation-based literature survey. *Omega*, 41(1), 3–15.

Mahalik, D. K., & Patel, G. (2010). Efficiency measurement using DEA and AHP: A case study on Indian ports. *IUP Journal of Supply Chain Management, 7*.

Mathivathanan, D., Govindan, K., & Haq, A.N. (2017). Explorando o impacto das capacidades dinâmicas no desempenho sustentável da empresa da cadeia de suprimentos usando o Processo hierárquico Grey Analytical. *Journal of Cleaner Production, 147*, 637–653.

Muchiri, P., & Pintelon, L. (2008). Performance measurement using overall equipment effectiveness (OEE): Literature review and practical application discussion. *International Journal of Production Research, 46*(13), 3517–3535.

Munneke, H. J., & Slade, B. A. (2001). A metropolitan transaction-based commercial price index: A time-varying parameter approach. *Real Estate Economics, 29*(1), 55–84.

Muthiah, K. M. N., & Huang, S. H. (2007). Overall throughput effectiveness (OTE) metric for factory-level performance monitoring and bottleneck detection. *International Journal of Production Research, 45*(20), 4753–4769.

Nakajima, S. (1989). *Introdução ao TPM – Total productive maintenance*. São Paulo: IMC Internacional Sistemas Educativos Ltda.

Neely, A., Gregory, M., & Platts, K. (1995). Performance measurement system design: A literature review and research agenda. *International Journal of Operations & Production Management, 15*(4), 80–116.

Oechsner, R., Pfeffer, M., Pfitzner, L., Binder, H., Müller, E., & Vonderstrass, T. (2002). From overall equipment efficiency (OEE) to overall Fab effectiveness (OFE). *Materials Science in Semiconductor Processing, 5*(4), 333–339.

Raouf, A. (1994). Improving capital productivity through maintenance. *International Journal of Operations & Production Management, 14*(7), 44–52.

Saaty, T. L. (1977). A scaling method for priorities in hierarchical structures. *Journal of Mathematical Psychology, 15*(3), 234–281.

Saaty, T. L. (1991). *Método de análise hierárquica*. Livro, São Paulo: Editora Makron.

Scott, D., & Pisa, R. (1998). Can overall factory effectiveness prolong Mooer's law? *Solid State Technology, 41*(3), 75–81.

Thanki, S., Govindan, K., & Thakkar, J. (2016). An investigation on lean-green implementation practices in Indian SMEs using analytical hierarchy process (AHP) approach. *Journal of Cleaner Production, 135*, 284–298.

5 Data Envelopment Analysis (DEA)

This chapter explains the concepts and foundations of DEA, which is the technique presented in this book, to analyze productivity and efficiency in production systems. We will discuss the concepts of DMU, the CRS and VRS models and the orientation types of the models. In addition, we will show how the targets and slack are stipulated, which are very useful concepts for managers. We also describe the types of technical efficiency which are calculated in DEA. We detail DEA equations in this chapter. To achieve a better understanding of DEA equations and of their concepts and fundamentals, illustrative examples are presented. We emphasize that these equations are inserted in an application that we have developed, which can be used to perform the calculations.

Based on the ideas of Farrell (1957) and the need to improve procedures for assessing the productivity and efficiency of production units, Charnes et al. (1978) developed a technique known as Data Envelopment Analysis (DEA), which is a non-parametric and frontier programming approach that is used to measure the efficiency of decision-making units (DMUs) that have multiple inputs and outputs (Charnes et al., 1978; Liu et al., 2013, 2016; Cook et al., 2014).

DEA enables overall and holistic assessments of a system's productivity and efficiency. A system can be understood as a complex arrangement of elements that operate with organized relations among them (Bertalanffy, 1975). A basic feature of a system is that it has inputs that are processed to generate the outputs. Inputs and outputs are known as system variables. Thus, the system considers variables that interrelate. In DEA, the system is called the DMU. Section 5.1 will explain in detail what a DMU is and how it can be formed.

The original idea of DEA was to provide a way that, within a set of Decision-Making Units (DMUs), it was possible to identify the most productive DMUs, presenting the best practices and forming an efficiency frontier. The efficiency frontier is the production curve that gathers together companies whose production performance is higher than the others. This considers criteria for solving a linear programming problem (PL) that takes into account the proportions of the quantities of inputs and products. Thus, the efficiency frontier refers to the points that demarcate the productivity at which a productive unit is technically efficient.

The efficiency of each DMU is defined as the ratio of the weighted sum of its products (outputs) to the weighted sum of the inputs needed to generate them. For each DMU considered to be inefficient, DEA identifies a set of references composed of one or more efficient DMUs that can be used as a reference for improvement (Cook & Seiford, 2009; Liu et al., 2013; Lee & Kim, 2014). This procedure is known as benchmarking.

The benchmarking performed in DEA makes it possible to assess whether or not a particular DMU is close to the most prominent DMUs, that is, those DMUs that are at the efficient frontier. Thus, the most prominent DMUs can provide important information that could make it possible to carryout direct actions towards the improvement of the inefficient DMUs.

The original DEA model was proposed by Charnes et al. (1978). In 1979, they adjusted it, and, in 1981, applied it in an empirical study they had conducted in US public schools. In general, a researcher's goal in using DEA is to minimize inputs and maximize output, in order to reduce input resources and increase the number of outputs from the system analyzed.

The choice between minimizing inputs and maximizing outputs is determined by the orientation of the model chosen. If the objective is to keep the outputs constant (e.g. the production volume of a certain product) and to determine the best use of the inputs in the process (e.g. raw materials), the model should be input oriented. However, if the objective is to keep resource consumption constant (e.g. raw materials) and maximize outputs (e.g. output volume of a particular item), the model should be output oriented. Input orientation is more normally used because inputs are regarded as more controllable. However, there are situations, such as increased market demand, where the use of the model orientation that maximizes output may be considered more interesting.

The context presented, however, may not be a general rule, with exceptions such as pollutants emitted from a production process. In this specific case, the higher the output level, the worse the performance of the process under analysis. Thus, there are DEA models that deal with such outputs (or inputs) which are considered undesirable (Seiford & Zhu, 2002). Considering, for example, a hospital efficiency study, the inputs could be the number of beds available and the hospital budget. The outcomes could be the number of patients attending and the number of trained nurses. With regard to the study of a university, for example, one could consider as inputs the quality of the students (percentage of students that obtain scholarships, their assessments, etc.). As outputs, one could consider the placement of students in internships and jobs (Cook et al., 2014).

5.1 DECISION-MAKING UNIT (DMU)

As regards a definition of DMU, this consists of a unit for decision-making of a productive unit. The term DMU can be defined as a comparative system. The possibility of comparing DMUs is a fundamental assumption that cannot be broken in the application of DEA. Examples of DMUs can be cited as projects, products, departments, divisions, administrative units or the company itself, among others. Thus, DMUs are considered units of analysis essential for use of data envelope analysis. Figure 5.1 illustrates a DMU in a production system, whether for goods or services. In addition to this, its association with the inputs and outputs used in DEA is also shown.

At the top of Figure 5.1, the generic model shows how analysis of productivity and efficiency of a system using DEA can be considered. In the central part of the model is the DMU. Thus, inserted in the DMU are the input variables, also called inputs, which are processed by the DMU itself. The result of this process generates

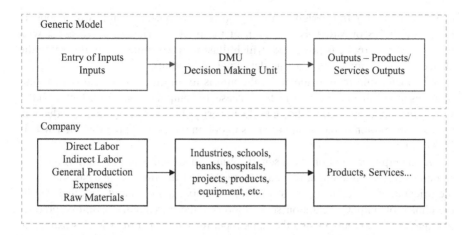

FIGURE 5.1   Relations among input, DMU and output – DEA.

the outputs that can be goods or services. In the lower part of Figure 5.1. an attempt is made to translate this model into some hypothetical cases, showing what a DMU can be, which can be the inputs or the outputs.

There are several models of DEA. Currently, the most widely used models are constant returns to scale (CRS) and variable returns to scale (VRS). We emphasize that the CRS and VRS models are mathematical models. The mathematical models are different from the generic model of analysis illustrated in Figure 5.1. These models are applied to various industrial and non-industrial contexts (banks, hospitals, education, etc.). The CRS and VRS model are presented in the following sections.

5.2   MODEL CRS (CONSTANT RETURNS TO SCALE)

The CRS model allows an objective assessment of the overall efficiency, identifying the value as a percentage of efficiency and consequently of inefficiency. Additionally, it allows identification of the sources or reasons for the inefficiencies of the DMU. This model operates with constant returns to scale and constructs a non-parametric linear surface enveloping the data. Thus, in the CRS model, there is a linear production function.

The CRS model should be used when there is a constant relationship on the scale between outputs and inputs of the DMUs being analyzed, for example, if an efficiency analysis among the world's largest automakers is being conducted. In this case, it is understood that companies operate on similar production scales. In another case, if an efficiency analysis among the world's largest consulting firms is being conducted, it is also understood that companies operate on a similar scale when providing a delivery service. In addition, in the two examples cited, the increase in inputs should generate an increase in outputs in the same proportion.

Normally, the CRS model is also recommended for performing internal analyses (benchmarks). This is because, in this case, a comparison is made of an organization's performance with itself over time, that is, with the same conditions and scale of production over the period analyzed. Another important point, after choosing the model (CRS or VRS), is definition of the model's orientation. To assist with this choice, Figure 5.2 is presented, which illustrates a demonstration regarding the choice of orientation of the CRS model (input and output).

When choosing input orientation, the aim is to minimize inputs, that is, to use input resources in an optimal way. In this case, the output remains constant. In other words, the input-orientation choice seeks to verify the amount of resources that can be saved in relation to the input variables to obtain the same output that the DMU is performing.

When output orientation is chosen, the aim is to keep inputs constant, using input resources in the same quantity as are being used at the time of analysis. However, the orientation to output refers to the goal of maximizing output. In other words, the choice of output orientation seeks to verify the maximum amount of production that can be executed using the same input resources that the DMU is using. An illustrative example of the input and output orientations of the CRS model is shown in Figure 5.3.

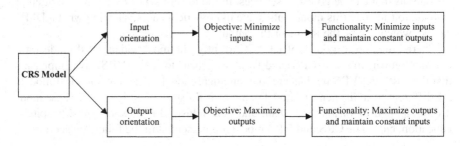

FIGURE 5.2   CRS model orientation.

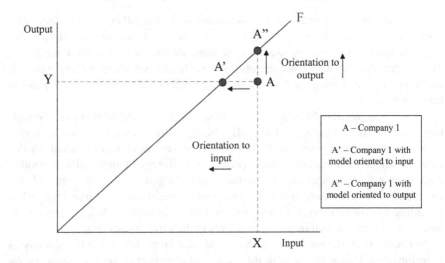

FIGURE 5.3   Representation of the CRS model oriented to input and output.

It can be seen that, for the CRS model, the frontier (F) is a straight 45° line. Efficiency (A), calculated by the output-oriented model, is equal to that calculated by the input-oriented model. In the case shown in Figure 5.3, the DMU has two options for becoming efficient. In the first option, the DMU should detect displacement of the inputs to the left, as shown by the indicative arrow (A to A'). In other words, this means that, to become efficient, the DMU needs to consume less resources to carry out the same production. In the second option, the DMU can search for upward shifts of outputs, as shown by the arrow (A to A"). Thus, to become more efficient the DMU needs to achieve more production using the same resources.

In the CRS model, the efficient DMUs are exclusively those that have a maximum productivity that, with the coefficients calculated by the mathematical model, is always equal to one. Section 5.2.1 shows the equations of the CRS model with input orientation.

5.2.1   CRS MODEL WITH INPUT ORIENTATION

The input-oriented CRS model is represented by Equations (5.1), (5.2) and (5.3):

$$\text{Max Eff}_0 = \frac{\sum_{j=1}^{m} u_j \, y_{j0}}{\sum_{i=1}^{n} v_i \, x_{i0}} \tag{5.1}$$

subject to:

$$\frac{\sum_{j=1}^{m} u_j \, y_{jk}}{\sum_{i=1}^{n} v_i \, x_{ik}} \leq 1, \ \forall k \tag{5.2}$$

$$uj \geq 0, \ \forall j$$
$$vi \geq 0, \ \forall i \tag{5.3}$$

where:
  Eff_0 = efficiency of DMU 0 under analysis
  u_j = weight calculated for output j, $j = 1, \ldots n$
  v_i = weight calculated for input i, $i = 1, \ldots n$
  y_{j0} = quantity of the output j for the DMU under analysis
  x_{i0} = quantity of the input i for the DMU under analysis
  y_{jk} = quantity of the output j for DMU k, $k = 1, \ldots n$
  x_{ik} = quantity of the input i for DMU k, $k = 1, \ldots n$
  k = number of DMUs under analysis
  m = number of outputs
  n = number of inputs

The mathematical model presented is fractional programming and should be solved for each of the DMUs. This mathematical model can also be transformed into a linear programming problem (PPL). For this, it is necessary that the denominator of the objective function be equal to a constant, normally equal to the unit (Mello et al., 2005). Such formulation is demonstrated in Equation 5.4, where the decision variables are the weights u_j and v_i.

$$\text{Max Eff}_0 = \sum_{j=1}^{m} u_j \, y_{j0}$$

subject to:

$$\sum_{i=1}^{n} v_i x_{i0} = 1$$

$$\sum_{j=1}^{m} u_j y_{jk} - \sum_{i=1}^{n} v_i x_{ik} \leq 0, \; \forall k$$

$$uj \geq 0, \; \forall j$$

$$vi \geq 0, \; \forall i$$

The first equation is the objective function (FO) of the mathematical programing model that should be maximized. The second equation is a set of constraints (one for each DMU assessed) that limits the productivity of all DMUs in the first equation. This restriction is very important because the productivity of a DMU can theoretically assume any value (it is unlimited) and, if it were not for this restriction, it would not be possible to maximize the objective function (Coelli et al., 2005). The third equation shows that the inputs used and the outputs generated should be greater than zero.

The input orientation, where efficiency is achieved through the reduction of inputs and resources, can best be visualized in the dual model, also known as the Envelope Model, as shown in 5.5.

$$\text{Min} \, h_0$$

subject to:

$$h_0 x_{j0} - \sum_{i=1}^{n} x_{ik} \, \lambda_k \geq 0, \; \forall i$$

$$-y_{j0} + \sum_{j=1}^{m} y_{jk} \, \lambda_k \geq 0, \; \forall j$$

$$\lambda_k \geq 0, \; \forall k \tag{5.5}$$

In Section 5.2.2, we show the equations of the CRS model with output orientation.

5.2.2 CRS MODEL WITH OUTPUT ORIENTATION

The equations of the output-oriented CRS model are (5.6), (5.7) and (5.8). It should be emphasized that the equations presented are mathematical models of fractional programing. The efficiency of the output-oriented models is calculated by the inverse of the objective function, i.e. efficiency (ho) = 1/E. This mathematical model defines the relationship of the inputs to the outputs.

$$Min\ h_0 = \frac{\sum_{i=1}^{n} v_i\, x_{i0}}{\sum_{j=1}^{m} u_j\, y_{j0}} \tag{5.6}$$

subject to:

$$\frac{\sum_{i=1}^{n} v_i x_{ik}}{\sum_{j=1}^{m} u_j y_{jk}} \leq 1,\ \forall k \tag{5.7}$$

$$uj \geq 0,\ \forall j$$
$$vi \geq 0,\ \forall i \tag{5.8}$$

where:
h_0 = 1/eff$_0$
v_i = weight calculated for the input i, i = 1, ...n
u_j = weight calculated for the output j, j = 1, ...n
x_{i0} = quantity of the input i for the DMU under analysis
y_{j0} = quantity of the output j for the DMU under analysis
x_{ik} = quantity of the input i for DMU k, k = 1, ...n
y_{jk} = quantity of the output j for DMU k, k = 1, ...n
k = number of DMUs under analysis
n = number of inputs
m = number of outputs

The transformation of the fractional programming model into a linear programming problem (PPL) is shown in Equation 5.9.

$$Min\ h_0 = \sum_{i=1}^{n} v_i x_{i0}$$

subject to:

$$\sum_{j=1}^{m} u_j y_{j0} = 1$$

$$\sum_{j=1}^{m} u_j y_{jk} - \sum_{i=1}^{n} v_i x_{ik} \leq 0,\ \forall k$$

$$u_j \geq 0, \ \forall j$$

$$v_i \geq 0, \ \forall i \tag{5.9}$$

The output orientation, in which the efficiency is reached with the maximization of the products can be better visualized in the dual model, which is also known as the Envelope Model, shown in Equation (5.10).

$$\text{Max } h_0$$

subject to:

$$x_{j0} - \sum_{i=1}^{n} x_{ik}\lambda_k \geq 0, \ \forall i$$

$$-h_0 \, y_{j0} + \sum_{j=1}^{m} y_{jk}\lambda_k \geq 0, \ \forall j$$

$$\lambda_k \geq 0, \ \forall k \tag{5.10}$$

For a better understanding of the functionalities of the mathematical models referring to the CRS model, an illustrative example of their application will be shown.

5.2.3 EXAMPLE OF AN ANALYSIS USING THE CRS MODEL (WITH INPUT ORIENTATION)

To illustrate the calculation of productivity and efficiency in the CRS model, the example elaborated by Macedo and Bengio (2003) is used. In this example, the authors discuss a snackbar network that has six units (branches) in different districts of a given city. The owner wanted to conduct a network analysis to check the productivity and efficiency of each snackbar individually. The snackbar chain offered only one standard meal consisting of a hamburger, fries and a soda.

When performing a productivity and efficiency analysis in this context, initially each of the six snackbars in the network was defined as a DMU. In addition, with the help of the owner, a survey of the inputs and outputs of the snackbars was carried out. The input data were: (i) daily expenditure on salaries, transformed into US$, and (ii) daily consumption of materials, in US$. The volume of daily sales of meals in units was considered to be the output. Data on the quantities of inputs and outputs are listed in Chart 5.1. Finally, it was defined that the analysis would be oriented to input, as the objective was to keep the outputs constant and to verify how much of each input could be saved in each of the DMUs considered to be inefficient.

For an understanding of this DEA example, the linear programming model will be used. It should be stressed that, if the fractional model is applied, it will also generate the same results. In our example, the PPL model for DMU1 will be

---

CHART 5.1

Data of Each DMU (Snackbar)

Snackbar DMUs	Salaries in US$ Input 1 (v1)	Materials in US$ Input 2 (v2)	Meals Sold in UN Output 1 (u1)
1	80	30	20
2	100	30	22
3	100	28	18
4	130	29	25
5	150	25	17
6	165	27	19

Source: Adapted from Macedo and Bengio (2003).

---

shown. The formulation of DMU1 will initially seek to establish the weights of each input.

$$\text{Max Eff}_1 = 20u1$$

subject to:

$$80v1 + 30v2 = 1$$
$$20u1 - 80v1 - 30v2 \leq 0$$
$$22u1 - 100v1 - 30v2 \leq 0$$
$$18u1 - 100v1 - 28v2 \leq 0$$
$$25u1 - 130v1 - 29v2 \leq 0$$
$$17u1 - 150v1 - 25v2 \leq 0$$
$$19u1 - 165v1 - 27v2 \leq 0$$

u1, v1, v2 ≥ 0

To obtain the final results, it is necessary to repeat this formulation for the other five DMUs, effecting the exchange of the objective function and values of the first row with the respective values of the outputs and inputs of each DMU. After execution of all the PPLs, the responses shown in Chart 5.2 are obtained.

The results show that DMUs 1 and 4 are efficient because their efficiency scores are equal to 1. In addition, it can be observed that DMUs 2, 3, 5 and 6 are inefficient because their efficiency scores are lower than 1. This can also be observed through the efficient frontier graph (Figure 5.4). In Figure 5.4 we used the values of output1/input1 on the X axis, and on the Y axis, the values of output1/input2.

From a managerial perspective, it becomes very useful to perform a benchmarking analysis, identifying which efficient DMU(s) (1 or 4) can be considered as a reference for inefficient DMUs. For this, the DEA technique provides a general weight, which represents how similar an inefficient DMU is to an efficient DMU. In benchmark

CHART 5.2

Efficiency of DMUs and Input and Output Weights

DMUs	Efficiency	Weights		
		Input 1 (v1)	Input 2 (v2)	Output 1 (u1)
1	1.00	0.0125	0.00	0.05
2	0.9931	0.004857	0.017143	0.045143
3	0.8414	0.005030	0.017751	0.046746
4	1.00	0.004304	0.015190	0.04
5	0.7888	0.00	0.04	0.0464
6	0.8162	0.00	0.037037	0.042963

Source: Adapted from Macedo, Bengio (2003).

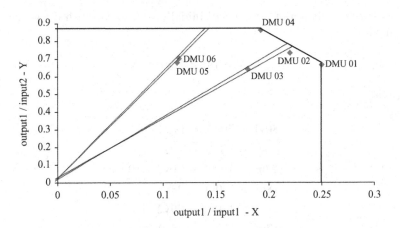

FIGURE 5.4   Efficient Frontier. (Adapted from Macedo and Bengio, 2003.)

language, it can be interpreted as follows. The DMU of greatest general weight is the one that should serve as reference for inefficient DMU to use as a basis for its practices, to become an efficient DMU. In other words, it can also be understood that the most efficient DMU with the greatest general weight is the reference that the inefficient DMU can "visit" to observe how it can improve its productive practices. Chart 5.3 shows the relationship between the inefficient DMUs and their references.

The values shown represent the relative weight associated with each efficient unit in calculating the efficiency rate for the inefficient units. This explains how much inefficient DMU inputs need to refer to the efficient DMU inputs (benchmark). With this reference, it is possible to establish goals so that the inefficient DMU can achieve efficiency by maintaining the current levels of outputs (considering that the model was input oriented). In DEA, this procedure is called target and slack analysis. The concepts of targets and slack will be explained and illustrated in Section 5.5.

After presentation and exemplification of the model with constant returns to scale (CRS), with orientations for input and output, the next section shows the variable returns model (VRS).

CHART 5.3
Benchmarking Analysis

	References	
Inefficient DMUs	Efficient DMUs	Weights
2	1	0.628571
	4	0.377143
3	1	0.394083
	4	0.404734
5	4	0.680000
6	4	0.760000

Source: Macedo & Bengio (2003).

5.3   MODEL VRS (VARIABLE RETURNS TO SCALE)

Banker et al. (1984) proposed the variable returns to scale model, known as VRS. Banker et al. (1984) argue that the VRS model is suitable for cases in which a DMU cannot be compared uniformly with all the DMUs of a given sector, but rather with the DMUs that operate on a scale similar to its own. Different from the CRS model, the production function in the VRS model is not linear and can be divided into two types of efficiency yields:

i) The first type is the model with a decreasing return to scale, in which an increase in the inputs causes a proportionally smaller increase in the outputs.

ii) The second type is the model with increasing returns to scale, where an increase in inputs causes a proportionally larger increase in outputs.

The VRS model assumes that maximum productivity varies according to production scale, so that it allows DMUs of completely different sizes to be compared and used in the same analysis. The only difference between CRS and VRS models is the addition of a variable "u" in the numerator, or a variable "v" in the denominator.

The variables "u" and "v" have the function of ensuring that the restrictions of the DMUs, which operate on a different scale from the DMU under analysis, do not limit their objective function. With this function it is possible to assess the return to scale with which the DMU is operating. If the value of "u" is greater than zero, it means that the company operates with decreasing returns to scale. If the value of "u" is less than zero, it means that the returns to scale are increasing. If the value of "u" equals zero, there are constant returns to scale. The variable "v" can also be used to estimate the type of DMU scale, but it should be interpreted in the opposite way to "u", that is, if v > 0, the returns will be increasing; if v = 0, the returns will be constant, and, if v < 0, the returns will be decreasing. Figure 5.5 provides an illustration of a variable return to scale model (VRS) case, comparing it with the CRS model.

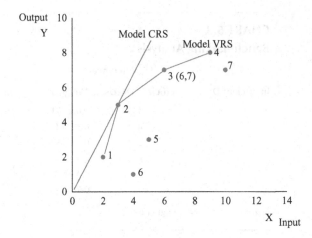

FIGURE 5.5 Exemplification of the VRS model compared to CRS. (Adapted from Cook and Seiford, 2009.)

In Figure 5.5, the line without the demarcated points represents the frontier of the CRS model, and the line with the demarcated points represents the frontier of the VRS model. The efficiency frontier of the CRS model is represented by a straight line while that of the VRS is not. The VRS model considers the frontier from point 1 to point 4. In this sense, point 2 is facing constant returns to scale ($v = 0$), whereas the points to the right of point 2 (2–3 and 3–4) have decreasing returns to scale ($v < 0$) (Cook & Seiford, 2009).

The orientation of the VRS model follows the same pattern as the CRS model. When input orientation is the option taken, the objective is to minimize inputs, that is, to use input resources in an optimal way. In this case, the output remains constant. In other words, the input-orientation choice seeks to verify what amount of resources can be saved in relation to the input variables to perform the same output that the DMU is performing. In addition, the linearization process of the VRS model uses the same procedures as the CRS model. However, the convexity of the VRS model is generated by including the variables "u" and "v".

When opting for output orientation, the goal is to keep inputs constant, using input resources in the same quantity that is being used at the time of analysis. However, the orientation to output refers to the quest to maximize the output. Thus, the choice of the output orientation seeks to verify the maximum amount of production that can be executed using the same input resources that the DMU is using. Figure 5.6 illustrates the input/output definition of the VRS model.

Section 5.3.1 shows the equations of the VRS model with input orientation.

$$\text{Max Eff}_0 = \frac{\sum_{j=1}^{m} u_j \, y_{j0} + u_0}{\sum_{i=1}^{n} v_i \, x_{i0}} \quad \text{or} \quad \text{Max Eff}_0 = \frac{\sum_{j=1}^{m} u_j \, y_{j0}}{\sum_{i=1}^{n} v_i \, x_{i0} + v_0} \tag{5.11}$$

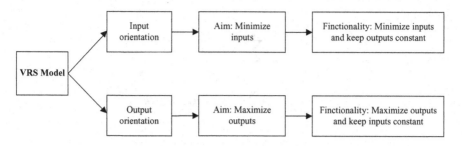

FIGURE 5.6   Orientations of the VRS model.

subject to:

$$\frac{\sum_{j=1}^{m} u_j\, y_{jk} + u_0}{\sum_{i=1}^{n} v_i\, x_{ik}} \le 1,\ \forall k \ \ \text{or}\ \ \frac{\sum_{j=1}^{m} u_j\, y_{jk}}{\sum_{i=1}^{n} v_i\, x_{ik} + v_0} \le 1,\ \forall k \tag{5.12}$$

$$u_j \ge 0,\ \forall j$$

$$v_i \ge 0,\ \forall i \tag{5.13}$$

where:

Eff_0 = efficiency of the DMU under analysis
u_j = weight calculated for the output j, j = 1, …n
v_i = weight calculated for the input i, i = 1, …n
y_{j0} = quantity of the output j for the DMU under analysis
x_{i0} = quantity of the input i for the DMU under analysis
y_{jk} = quantity of the output j for DMU k, k = 1, …n
x_{ik} = quantity of the input i for DMU k, k = 1, …n
u = variable of return to scale of the numerator
v = variable of return to scale of the denominator
k = number of DMUs under analysis
m = number of outputs
n = number of inputs

The transformation of the fractional programming model into a linear programming problem (PPL) is shown in Equation (5.14).

5.3.1   VRS MODEL WITH INPUT ORIENTATION

The equations of the VRS model with input orientation are represented by Equations (5.11), (5.12) and (5.13):

$$\text{Max Eff}_0 = \sum_{j=1}^{m} u_j\, y_{j0} + u_0$$

subject to:

$$\sum_{i=1}^{n} v_i x_{i0} = 1$$

$$-\sum_{i=1}^{n} v_i\, x_{ik} + \sum_{j=1}^{m} u_j\, y_{jk} + u_0 \leq 0, \ \ \forall k$$

$$uj \geq 0, \ \mathrm{u} \in \mathfrak{R}$$

$$vi \geq 0, \ \mathrm{u} \in \mathfrak{R} \tag{5.14}$$

The input orientation, in which efficiency is achieved with a reduction of inputs and resources, can best be visualized in the dual model, shown in 5.15.

$$\mathrm{Min}\, h_0$$

subject to:

$$h_0 x_{i0} - \sum_{i=1}^{n} x_{ik} \lambda_k \geq 0, \ \ \forall i$$

$$-y_{j0} + \sum_{j=1}^{m} y_{jk} \lambda_k \geq 0, \ \ \forall j$$

$$\sum_{j=1}^{m} \lambda_k = 1$$

$$\lambda_k \geq 0, \ \ \forall k \tag{5.15}$$

Section 5.3.2 shows the equations of the VRS model with orientation to output.

5.3.2 VRS MODEL WITH OUTPUT ORIENTATION

The equations of the VRS model with output orientation are represented by Equations (5.16), (5.17) and (5.18):

$$\mathrm{Min\ Eff}_0 = \frac{\sum_{i=1}^{n} v_i\, x_{i0} + v_0}{\sum_{j=1}^{m} u_j\, y_{j0}} \quad \text{or} \quad \mathrm{Min\ Eff}_0 = \frac{\sum_{i=1}^{n} v_i\, x_{i0}}{\sum_{j=1}^{m} u_j\, y_{j0} + u_0} \tag{5.16}$$

subject to:

$$\frac{\sum_{i=1}^{n} v_i\, x_{ik} + v_0}{\sum_{j=1}^{m} u_j\, y_{jk}} \leq 1, \ \ \forall k \quad \text{or} \quad \frac{\sum_{i=1}^{n} v_i\, x_{ik+}}{\sum_{j=1}^{m} u_j\, y_{jk} + u_0} \leq 1, \ \ \forall k \tag{5.17}$$

$$uj \geq 0, \ \forall j$$

$$vi \geq 0, \ \forall i \tag{5.18}$$

where:

Eff$_0$ = efficiency of the DMU under analysis
v_i = weight calculated for the input i, i=1, ...n
u_j = weight calculated for the output j, j=1, ...n
x_{i0} = quantity of the input i for the DMU under analysis
y_{j0} = quantity of the output j for the DMU under analysis
x_{ik} = quantity of the input i for DMU k, k=1, ...n
y_{jk} = quantity of the output j for DMU k, k=1, ...n
u = variable of the return to scale of the numerator
v = variable of the return to scale of the denominator
k = number of DMUs under analysis
n   = number of inputs
m   = number of outputs

The transformation of the fractional programming model into linear programming problem (PPL) is shown in Equation 5.19.

$$\text{Min Eff}_0 = \sum_{i=1}^{n} v_i x_{i0} + v_0$$

subject to:

$$\sum_{j=1}^{m} u_j y_{j0} = 1$$

$$-\sum_{i=1}^{n} v_i x_{ik} + \sum_{j=1}^{m} u_j y_{jk} + v_0 \leq 0, \ \forall k$$

$$uj \geq 0, \ u \in \Re$$

$$vi \geq 0, \ u \in \Re \tag{5.19}$$

The output orientation, where efficiency is achieved with maximization of the products, can better be visualized in the dual model, shown in 5.20.

$$\text{Max } h_0$$

subject to:

$$x_{i0} - \sum_{i=1}^{n} x_{ik} \lambda_k \geq 0, \ \forall i$$

$$-h_0\, y_{j0} + \sum_{j=1}^{m} y_{jk}\lambda_k \geq 0, \ \forall j$$

$$\sum_{j=1}^{m} \lambda_k = 1$$

$$\lambda_k \geq 0, \ \forall k \tag{5.20}$$

For a better understanding of the functionalities of the mathematical models related to the VRS model, an illustrative example of application will be shown.

5.3.3   EXAMPLE OF AN ANALYSIS USING THE VRS MODEL (WITH INPUT ORIENTATION)

To illustrate the way of calculating productivity and efficiency in the VRS model, we use the same example elaborated by Macedo and Bengio (2003). In this case, the same inputs (daily expenditure with salary in US\$ and daily consumption of materials in US\$) and output (daily volume of meals) will be used in the same DMUs (snackbars). In addition, the same model orientation (input) will be used. Thus, the formulation of the linear programming model of the DMU1 will seek to establish the weights of each input:

$$MAX_{eff_1} = 20u1 + u0$$

subject to:

$$80v1 + 30v2 = 1$$
$$20u1 + u - 80v1 - 30v2 \leq 0$$
$$22u1 + u - 100v1 - 30v2 \leq 0$$
$$18u1 + u - 100v1 - 28v2 \leq 0$$
$$25u1 + u - 130v1 - 29v2 \leq 0$$
$$17u1 + u - 150v1 - 25v2 \leq 0$$
$$19u1 + u - 165v1 - 27v20$$

$$u1, v1, v2 \geq 0, \ \mathbf{u} \in \Re$$

In this analysis it is also necessary to repeat this formulation for the other five DMUs, effecting the exchange of the objective function and values of the first row with the respective values of the outputs and inputs foreach DMU. In addition, the variable "u" is added. After the execution of all the PPLs, the following responses are obtained (Chart 5.4).

Unlike the analysis in the CRS model, in the VRS model the DMUs 1, 2, 3, 4 and 5 are efficient and only DMU 6 is inefficient. Thus, Chart 5.5 shows the relationship of DMU 6 with its reference DMUs.

CHART 5.4

Efficiency of the DMUs and Weights of the Outputs and Inputs

		Weights			
DMUs	Efficiency	Input 1 (v1)	Input 2 (v2)	Output 1 (u1)	u0
1	1.00	0.0125	0	0.05	0
2	1.00	0.01	0	0.1	−1.2
3	1.00	0.00176471	0.02941176	0.01176471	0.78823529
4	1.00	0.0043038	0.01518987	0.04	0
5	1.00	0	0.04	0.02	0.66
6	0.962963	0	0.03703704	0.01851852	0.61111111

CHART 5.5

Benchmarking Analysis

	References	
Inefficient DMUs	Efficient DMUs	Weights
6	4	0.25
	5	0.75

With this reference, it is possible to establish goals so that the inefficient DMU (6) can achieve efficiency by maintaining the current output levels (considering that the model was input oriented).

5.4 RELATION OF THE CRS AND VRS MODELS FOR CALCULATION OF THE EFFICIENCY OF SCALE

In the analysis of efficiency with DEA, it is possible to make a relation between the efficiencies calculated by the CRS and VRS models to obtain an efficiency of scale. It is to be remembered that the efficiency of scale is the result of the maximum production level located under the efficient frontier, which consists of an optimal unit (DMU) of operation where the reduction or increase in the scale of production implies a reduction in efficiency.

Scale efficiency can be obtained by calculating the ratio between the efficiency with constant returns (CRS) and the efficiency with variable returns (VRS), as demonstrated in Equation 5.21.

$$\text{Efficiency of scale} = \frac{\text{CRS Efficiency}}{\text{VRS Efficiency}} \tag{5.21}$$

Calculation of the efficiency of scale is presented for the example of the snackbars, the results of which are shown in Chart 5.6.

CHART 5.6

Examples of Calculation of the Efficiency of Scale

DMU	CRS Efficiency	VRS Efficiency	Efficiency of scale
1	1.00	1.00	1.00
2	0.9931	1.00	0.9931
3	0.8414	1.00	0.8414
4	1.00	1.00	1.00
5	0.7888	1.00	0.7888
6	0.8162	0.962963	0.8476

When the scale efficiency value is equal to one, it indicates that the DMU operates on its maximum production scale, also called the optimal production scale. When the result of the scaling efficiency is different from one, it is necessary to calculate the scale yield performance using Equation 5.22. If the sum of the calculated weights (λ) of the reference DMUs (benchmark) is greater than one, the returns to scale will be increasing. If it is smaller than one, the returns to scale will be decreasing (Banker et al., 1984).

$$\sum_{k=1}^{n} \lambda_k = 1, \text{Constant Returns}$$

$$\sum_{k=1}^{n} \lambda_k > 1, \text{Increasing Returns}$$

$$\sum_{k=1}^{n} \lambda_k < 1, \text{Decreasing Returns}$$

$$\lambda \geq 0, \ \forall k \tag{5.22}$$

where:

λ_k = the sum of the calculated DMU weights considered as reference (benchmark).

Calculation of the performance of scale is presented for the example of the snack-bars for illustration, the results of which are shown in Chart 5.7.

The column "Demonstration of the calculation" presents the sum of the weights calculated in the reference DMUs to obtain the benchmarking of the inefficient DMUs. These results were presented previously in Charts 5.3 and 5.5. Finally, the scale performance of each DMU is defined. In the next section, the concepts related to the analysis of the targets and slack are presented.

5.5 ANALYSIS OF TARGETS AND SLACK

The efficiency analyses using DEA provide calculation of the targets and slack (associated waste), which are the reference values that should be used to establish

---

CHART 5.7
Examples of the Calculation of the Performance of Scale

DMU	Efficiency of scale	Demonstration of the calculation	Performance of scale
1	1.00	=1	Constant return
2	0.9931	0.628571 + 0.377143 = 1.0057	Increasing return
3	0.8414	0.394083 + 0.404734 = 0.7989	Decreasing return
4	1.00	=1	Constant return
5	0.7888	0.68 = 0.68	Decreasing return
6	0.8476	0.25 + 0.75 + 0.76 = 1.7600	Increasing return

---

improvement goals for inefficient DMUs in relation to each input or output used in the calculation. Targets and slack provide indications that resources are being underused in the process under review. By this functionality, the targets and slack are very important parameters for managers, since they can provide a measure of reduction in the use of the inputs or for the increase in the production of the production systems.

If the parameters suggested by the targets and slack have been reached, an inefficient DMU can become efficient. In other words, the targets and slack provided by DEA guide the managers to the measures that can be taken to improve the efficiency of the system. In our opinion, this is one of the main reasons that makes the DEA technique attractive to companies. There is no other technique for analyzing productivity and efficiency that provides this information. To calculate the target (Equation 5.23) for a given DMU, it is necessary to perform the product of the current position of an input by the calculated weights (λ) belonging to the reference DMU, i.e. the DMU that is the benchmark.

$$\sum_{k=1}^{n} x_{ik} \, \lambda_k \qquad (5.23)$$

where:

x_{ik} = quantity of the input i for DMU k, $k = 1, \ldots n$
λ_k = sum of the weights calculated for the DMUs considered to be the benchmark

For a better understanding, we will show the calculation of the targets and slack for the same example of the productivity and efficiency analysis of the snackbars used in Section 5.2.3. Chart 5.8 shows the targets and slack for each input of the inefficient DMUs.

Using the CRS model, in the case of DMU 6, the following interpretation can be made. In order to produce and sell 19 snacks (output) daily, the snackbar spends US$ 165 on salaries and US$ 27 on materials. However, if this DMU is as efficient as its benchmark DMU, which is DMU 4, it can produce the same 19 snacks per day, spending US$ 129.24 per day on salaries, i.e. US$ 35.76 less than is currently spent. In addition, it can spend U$ 22.04 on raw material per day, or U$ 4.96 less than is currently spent on materials. For the other DMUs (2, 3 and 5), the same interpretation can be made.

CHART 5.8
Levels of Target Inputs and Associated Waste (Slack)

DMU 2

Inputs	Reference Set		Input Target	Current Input	Slack of Inputs
	DMU 1	DMU 4			
Input 1 (US$)	80 × 0.628571	+ 130 × 0.377143	99.31427	100	0.68573
Input 2 (US$)	30 × 0,628571	+ 29 × 0.377143	29.794277	30	0.205723

DMU 3

Inputs	Reference Set		Input Target	Current Input	Slack of Inputs
	DMU 1	DMU 4			
Input 1 (US$)	80 × 0.394083	+ 130 × 0.404734	84.14205	100	15.85794
Input 2 (US$)	30 × 0.394083	+ 29 × 0.404734	23.559775	28	4.440224

DMU 5

Inputs	Reference Set	Input Target	Current Input	Slack of Inputs
	DMU 4			
Input 1 (US$)	130 × 0.68	88.40	150	61.60
Input 2 (US$)	29 × 0.68	19.72	25	5.28

DMU 6

Inputs	Reference Set	Input Target	Current Input	Slack of Inputs
	DMU 4			
Input 1 (US$)	130 × 0.76	129.24	165	35.76
Input 2 (US$)	29 × 0.76	22.04	27	4.96

If the basis of the analysis is the efficiency calculated by the VRS model, the results for the targets and slack are calculated only for DMU 6, since this is the only DMU considered to be inefficient. The data are shown in Chart 5.9.

Thus, with the use of the VRS model, in the case of DMU 6, the following interpretation can be made. To produce 19 snacks a day (output), the snackbar spends U$ 165.00 on salaries and U$ 27.00 on materials per day. However, if this DMU is as efficient as its benchmarks, that is, DMUs 4 and 5, it can produce the same 19 snacks per day, spending U$ 145.00 per day on salaries, U$ 20.00 less than it currently spends. In addition, U$ 26.00 can be spent on raw material per day, or U$ 1.00 less than is currently spent.

5.6 TYPES OF TECHNICAL EFFICIENCY CALCULATED IN DATA ENVELOPMENT ANALYSIS (DEA)

The DEA technique allows calculation of different types of technical efficiency: i) standard; ii) inverted frontier; iii) composite; and iv) composite* (normalized). The relationship between these types of efficiency is highlighted in Figure 5.7. However,

---

CHART 5.9

Ideal Input Levels and Associated Waste

DMU 6

Inputs	Reference Set		Input Target	Current Input	Slack of Inputs
	DMU 4	DMU 5			
Input 1 (US$)	130 × 0.25	+ 150 × 0.75	145	165	20.00
Input 2 (US$)	29 × 0.25	+ 25 × 0.75	26	27	1.00

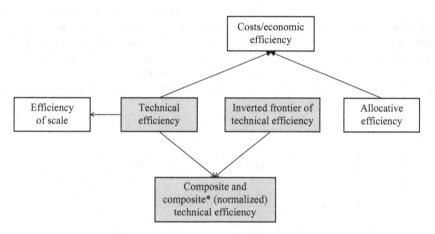

FIGURE 5.7　Relation between types of efficiency in DEA.

we highlight that, in the explanations and examples shown in the book, we focus on the calculation of technical efficiency (also considered standard).

From the calculation of the technical efficiency (standard), it becomes possible to calculate the inverted frontier of the composite and (normalized) composite technical efficiency. It should be noted that the calculation of these technical efficiencies (inverted, composite and composite*) is possible in CRS and VRS models. The concepts regarding each type of technical efficiency are described in Chart 5.10.

The technical efficiency is calculated as shown previously. The inverted frontier of technical efficiency can be seen as a pessimistic assessment of the DMUs, and its concept was introduced by Yamada et al. (1994). However, it became more representative when addressed by Entani et al., (2002), who named it IDEA (DEA inefficiency).

The inverted frontier of technical efficiency, or IDEA, assesses the inefficiency of a DMU by constructing a frontier made up of units with the worst managerial practices, called the inefficient frontier (Entani et al., 2002; Silveira et al., 2012). In order to calculate the frontier of inefficiency, an exchange is made of the inputs with the outputs and vice versa, of the original DEA model. Figure 5.8 shows the two frontiers, DEA and the inverted IDEA, for an illustrative DEA case.

CHART 5.10
Types of Efficiency Calculated in DEA

Type of efficiency	Description
Technical efficiency (standard)	Constituted of efficient units, that is, DMUs with the best performance, which execute the best practices
Inverted frontier of technical efficiency	Constituted of inefficient units, that is, DMUs with the worst performance, which do not execute the best practices
Composite technical efficiency	Constituted of an aggregate index between technical efficiency (standard) and inverted frontier of technical efficiency. For a DMU to achieve the best performance, it is necessary to obtain a high score in technical efficiency (standard) and a lower score on the inverted frontier of technical efficiency
Composite technical efficiency* (normalized)	Constituted of normalization of the efficiency score of the DMU with better performance in the composite technical efficiency. In this case, this DMU with better performance is considered to be 100% efficient and the efficiency scores are normalized successively

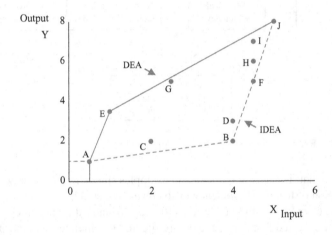

FIGURE 5.8 Technical (standard) and inverted frontier/IDEA. (Adapted from Entani et al., 2002.)

The assessment of the inverted frontier of the technical efficiency allows calculation of the composite technical efficiency. The composite technical efficiency can be used to circumvent a problem commonly faced by DEA users, which is the problem of discrimination between efficiency indexes. According to Adler and Yazhemsky (2010), if a DEA model has a relatively high number of variables (inputs and outputs) in relation to the defined DMUs, it will tend to present a discrimination problem among the DMUs. The problem of discrimination occurs when it is not possible to distinguish efficient DMUs from inefficient ones, prejudicing analysis of results. It can be noticed that an analysis has a discrimination problem when all the DMUs have efficiency equal to or very close to 1.

To overcome this problem, a number of suggestions have been proposed. Golany and Roll (1989) suggested that the number of DMUs defined should be at least double the combined number of inputs and outputs, whereas Banker et al. (1989) proposed that the number of DMUs should be at least three times the combined number of inputs and outputs. However, even following these rules, there is no guarantee that the problem of discrimination will be solved.

Composite technical efficiency was developed by Brazilian researchers (Mello et al., 2008), and helps to solve the discrimination problem. Composite technical efficiency is an aggregate index, which corresponds to the arithmetic mean between the technical efficiency in relation to the standard frontier and the technical inefficiency in relation to the inverted frontier (Mello et al., 2008; Gilsa et al., 2017), according to Equation 5.24.

$$\text{Composite Technical Efficiency} = \frac{\text{Technical Efficiency} + \left(1 - \text{Inverted Frontier}\right)}{2}$$

(5.24)

For a DMU to have maximum composite technical efficiency, it needs to perform well at the standard frontier and not perform well at the inverted frontier (Mello et al., 2008; Gilsa et al., 2017). Furthermore, this composite technical efficiency index can be presented in a standardized way.

For the example of the snackbars, the results of the calculation of the types of technical efficiency are presented in Chart 5.11. We emphasize that, for this illustrative example, we use the values related to the efficiencies obtained with the calculations of the CRS model, as described in Section 5.2. 3.

5.7 FINAL CONSIDERATIONS ON DEA

Since its development in 1978, DEA has been gaining increasing currency from researchers and managers, who have sought to apply it in disparate branches of the economy, business and society in general. Liu et al. (2013) developed a study and proved that data envelopment analysis is widely accepted in the academic and business fields and is being continually developed by researchers. The authors (Liu et al., 2013) identified the main works that contributed to the evolution of DEA, and these are listed in Chart 5.12.

CHART 5.11

Exemplification of the Calculation of the Types of Technical Efficiency

DMU	Technical efficiency (standard)	Inverted frontier of the technical efficiency	Composite technical efficiency	Composite technical efficiency* (normalized)
1	1.00	0.9642	0.5178	0.8365
2	0.9931	0.8766	0.5582	0.9017
3	0.8414	1.00	0.4207	0.6795
4	1.00	0.7618	0.6190	1.00
5	0.7888	1.00	0.3944	0.6370
6	0.8162	0.9842	0.4160	0.6720

CHART 5.12
Major Articles About DEA

Author(s)/Year	Publication title	Periodical
Charnes et al. (1978)	Measuring the efficiency of decision-making units	European Journal of Operational Research
Charnes et al. (1979)	Measuring the efficiency of decision-making units	European Journal of Operational Research
Charnes et al. (1981)	Assessing program and managerial efficiency: an application of data envelopment analysis to program follow through.	Management Science
Banker et al. (1984)	Some models for estimating technical and scale inefficiencies in data envelopment analysis	Management Science
Charnes et al. (1985)	Foundations of data envelopment analysis for Pareto-Koopmans efficient empirical production functions	Journal of Econometrics
Charnes et al. (1986)	Classifying and characterizing efficiencies and inefficiencies in data development analysis	Operations Research Letters
Seiford and Thrall (1990)	Recent developments in DEA: the mathematical programming approach to frontier analysis	Journal of Econometrics
Thompson et al. (1990)	The role of multiplier bounds in efficiency analysis with application to Kansas farming	Journal of Econometrics
Andersen and Petersen (1993)	A procedure for ranking efficient units in data envelopment analysis	Management Science
Athanassopoulos and Ballantine (1995)	Ratio and frontier analysis for assessing corporate performance: evidence from the grocery industry in the UK	Journal of the Operational Research Society
Athanassopoulos (1995)	Performance improvement decision aid systems (PIDAS) in retailing organizations using data envelopment analysis	Journal of Productivity Analysis
Seiford (1996)	Data envelopment analysis: the evolution of the state of the art (1978–1995)	Journal of Productivity Analysis
Kneip et al. (1998)	A note on the convergence of non-parametric DEA estimators for production efficiency scores	Econometric Theory
Simar and Wilson (1999)	Some problems with the Ferrier/Hirschberg bootstrap idea	Journal of Productivity Analysis
Simar and Wilson (2000)	Statistical inference in non-parametric frontier models: The state of the art	Journal of Productivity analysis
Fried et al. (2002)	Accounting for environmental effects and statistical noise in data envelopment analysis	Journal of Productivity analysis
Simar and Wilson (2007)	Estimation and inference in two-stage, semi-parametric models of production processes	Journal of Econometrics
Banker and Natarajan (2008)	Assessing contextual variables affecting productivity using data envelopment analysis	Operations Research
McDonald (2009)	Using least squares and Tobit in second stage DEA efficiency analyses	European Journal of Operational Research

Source: Based on Liu et al. (2013).

Liu et al. (2013) considered these 19 articles to be the main published works on DEA over the years. The main one, as expected, is the seminal work of Charnes et al. (1978), since the article establishes the bases of the DEA methodology. Subsequently, Charnes et al. (1979) wrote a one-page note, modifying the constraints of the main formulation of the 1978 paper. In their subsequent article, Charnes et al. (1981) applied the methodology to assess public education programs in the United States. The Banker et al. (1984) article proposed the variable return scale model. Charnes et al. (1985) presented an additive model and established the connection of DEA with the production theory through analysis of Pareto-Koopmans capacities. Charnes et al. (1986) characterized and classified the efficiencies and inefficiencies in DEA.

Later works advanced the discussion about DEA (Seiford & Thrall, 1990; Andersen & Petersen, 1993; Seiford, 1996; Kneip et al., 1998; Simar & Wilson, 1999, 2000; Fried et al., 2002). Applications were published on the agricultural sector (Thompson et al., 1990) and in retailing (Athanassopoulos & Ballantine, 1995; Athanassopoulos, 1995). Some authors (Simar & Wilson, 2007; Banker & Natarajan, 2008; McDonald, 2009) provided a statistical basis for the two-stage approach to analysis. The two-stage approach consists of using DEA to calculate efficiency in stage 1 and using other techniques to relate the influences to the scores of the efficiency factors observed (McDonald, 2009).

The research carried out by Paiva Junior (2000) identified several areas that use DEA for performance assessment, either with the usual objective of determining efficiency, or in order to help agents acting in the choice of alternative actions. These areas of DEA use were grouped, as shown in Figure 5.9, but without addressing the objectives of these applications.

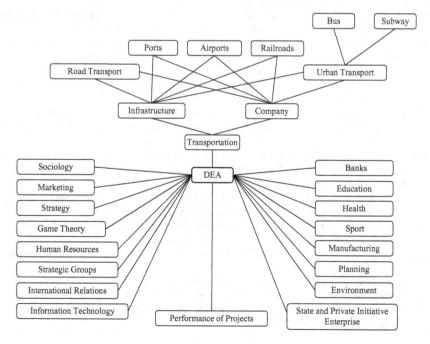

FIGURE 5.9 DEA application areas. (Adapted from Paiva Junior, 2000.)

Thus, we emphasize that DEA applications can be useful to solve problems, such as the measurement of the results obtained in relation to the resources used in the production processes. Moreover, this measurement can be carried out in several areas and different fields of knowledge, showing the potential of the DEA technique.

REFERENCES

Adler, N., & Yazhemsky, E. (2010). Improving discrimination in data envelopment analysis: PCA–DEA or variable reduction. *European Journal of Operational Research, 202*(1), 273–284.

Andersen, P., & Petersen, N. C. (1993). A procedure for ranking efficient units in data envelopment analysis. *Management Science, 39*(10), 1261–1264.

Athanassopoulos, A. D. (1995). Performance improvement decision aid systems (PIDAS) in retailing organizations using data envelopment analysis. *Journal of Productivity Analysis, 6*(2), 153–170.

Athanassopoulos, A. D., & Ballantine, J. A. (1995). Ratio and frontier analysis for assessing corporate performance: Evidence from the grocery industry in the UK. *Journal of the Operational Research Society, 46*(4), 427–440.

Banker, R. D., & Natarajan, R. (2008). Evaluating contextual variables affecting productivity using data envelopment analysis. *Operations Research, 56*(1), 48–58.

Banker, R. D., Charnes, A., & Cooper, W. W. (1984). Some models for estimating technical and scale inefficiencies in data envelopment analysis. *Management Science, 30*(9), 1078–1092.

Banker, R. D., Charnes, A., Cooper, W. W., Swarts, J., & Thomas, D. A. (1989). An introduction to data envelopment analysis with some of its models and their uses. *Research in Governmental and Nonprofit Accounting, 5*, 125–163.

Bertalanffy, L. (1975). *Teoria geral dos sistemas.* Vozes.

Charnes, A., Cooper, W. W., & Rhodes, E. (1978). Measuring the efficiency of decision-making units. *European Journal of Operational Research, 2*(6), 429–444.

Charnes, A., Cooper, W. W., & Rhodes, E. (1979). Measuring the efficiency of decision-making units. *European Journal of Operational Research, 3*(4), 339.

Charnes, A., Cooper, W. W., & Rhodes, E. (1981). Evaluating program and managerial efficiency: An application of data envelopment analysis to program follow through. *Management Science, 27*(6), 668–697.

Charnes, A., Cooper, W. W., & Thrall, R. M. (1986). Classifying and characterizing efficiencies and inefficiencies in data development analysis. *Operations Research Letters, 5*(3), 105–110.

Charnes, A., Cooper, W. W., Golany, B., Seiford, L., & Stutz, J. (1985). Foundations of data envelopment analysis for Pareto-Koopmans efficient empirical production functions. *Journal of Econometrics, 30*(12), 91–107.

Coelli, T. J., Rao, D. S. P., O'Donnell, C. J., & Battese, G. E. (2005). *An introduction to efficiency and productivity analysis.* (2nd ed., p. 349). New York: Springer.

Cook, W. D., & Seiford, L. M. (2009). Data envelopment analysis (DEA)–Thirty years on. *European Journal of Operational Research, 192*(1), 1–17.

Cook, W. D., Tone, K., & Zhu, J. (2014). Data envelopment analysis: Prior to choosing a model. *Omega, 44*, 1–4.

Entani, T., Maeda, Y., & Tanaka, H. (2002). Dual models of interval DEA and its extension to interval data. *European Journal of Operational Research, 136*(1), 32–45.

Farrell, M. J. (1957). The measurement of productive efficiency. *Journal of the Royal Statistical Society, 120*(3), 253–290.

Fried, H. O., Lovell, C. K., Schmidt, S. S., & Yaisawarng, S. (2002). Accounting for environmental effects and statistical noise in data envelopment analysis. *Journal of productivity Analysis*, *17*(1), 157–174.

Gilsa, C. V., Lacerda, D. P., Lacerda, D. P., Camargo, L. F. R., Camargo, L. F. R., & Cassel, R. A. (2017). Longitudinal evaluation of efficiency in a petrochemical company. *Benchmarking: An International Journal*, *24*(7), 1786–1813.

Golany, B., & Roll, Y. (1989). An application procedure for DEA. *Omega*, *17*(3), 237–250.

Kneip, A., Park, B. U., & Simar, L. (1998). A note on the convergence of nonparametric DEA estimators for production efficiency scores. *Econometric Theory*, *14*(6), 783–793.

Lee, H., & Kim, C. (2014). Benchmarking of service quality with data envelopment analysis. *Expert Systems with Applications*, *41*(8), 3761–3768.

Liu, J. S., Lu, L. Y., & Lu, W. M. (2016). Research fronts in data envelopment analysis. *Omega*, *58*, 33–45.

Liu, J. S., Lu, L. Y., Lu, W. M., & Lin, B. J. (2013). Data envelopment analysis 1978–2010: A citation-based literature survey. *Omega*, *41*(1), 3–15.

Macedo, M. A., & Bengio, M. C. (2003). Avaliação de eficiência organizacional através de análise envoltória de dados. *VIII Congresso3 Internacional de Costos*, 26.

McDonald, J. (2009). Using least squares and Tobit in second stage DEA efficiency analyses. *European Journal of Operational Research*, *197*(2), 792–798.

Mello, J. C. C. B. S., Meza, L. A, Gomes, E. G., & Neto, L. B. (2005). Curso de análise envoltória de dados. *XXXVII Congresso Brasileiro de Pesquisa Operacional*, 2521–2547.

Mello, J. C. C. B. S., Gomes, E. G., Meza, L. A., & Leta, F. R. (2008). DEA advanced models for geometric evaluation of used lathes. *WSEAS Transactions on Systems*, *7*(5), 510–520.

Paiva Junior, H. D. (2000). *Avaliação de desempenho de ferrovias utilizando a abordagem integrada DEA/AHP*.

Seiford, L. M. (1996). Data envelopment analysis: The evolution of the state of the art (1978–1995). *Journal of Productivity Analysis*, *7*(23), 99–137.

Seiford, L. M., & Thrall, R. M. (1990). Recent developments in DEA: The mathematical programming approach to frontier analysis. *Journal of Econometrics*, *46*(12), 7–38.

Seiford, L. M., & Zhu, J. (2002). Modeling undesirable factors in efficiency evaluation. *European Journal of Operational Research*, *142*(1), 16–20.

Silveira, J. Q., Meza, L. A., & Mello, J. S. (2012). Identificação de benchmarks e anti-benchmarks para companhias aéreas usando modelos DEA e fronteira invertida. *Production*, *22*(4), 788–795.

Simar, L., & Wilson, P. W. (1999). Some problems with the Ferrier/Hirschberg bootstrap idea. *Journal of Productivity Analysis*, *11*(1), 67–80.

Simar, L., & Wilson, P. W. (2000). Statistical inference in nonparametric frontier models: The state of the art. *Journal of Productivity Analysis*, *13*(1), 49–78.

Simar, L., & Wilson, P. W. (2007). Estimation and inference in two-stage, semi-parametric models of production processes. *Journal of Econometrics*, *136*(1), 31–64.

Thompson, R. G., Langemeier, L. N., Lee, C. T., Lee, E., & Thrall, R. M. (1990). The role of multiplier bounds in efficiency analysis with application to Kansas farming. *Journal of Econometrics*, *46*(12), 93–108.

Yamada, Y., Matui, T., & Sugiyama, M (1994). New analysis of efficiency based on DEA. *Journal of the Operations Research Society of Japan*, *37*, 158–167.

6 Modeling Method in DEA (MMDEA)

This chapter presents a modeling method to guide those interested in using the DEA technique. The proposed method is called DEA Modeling Method in DEA (MMDEA) and was developed by the authors of this book. MMDEA has been used in practical situations, contributing to the development of scientific research and managerial analyses on productivity and efficiency in organizations. Each step of the MMDEA is explained in detail to make it easier to understand. We hope it will be useful to readers in carrying out future analyses using DEA.

6.1 MODEL DEVELOPMENT – CONCEPTS AND IMPORTANCE

A model is an external explicit representation of part of the reality from the perspective of a person who wants to use the model to understand, change, manage and control part of this reality (Pidd, 1996). In general, models are used to describe how a system works or behaves. Thus, models can generate results that, when interpreted, can aid managers' decision-making process.

Construction of a model requires a rigorous process. This is because the structure of a model includes other structures, such as inventories, personnel, materials and money, among others, which are characteristics of a system (Sterman et al., 2015). Because of this, in almost all modeling processes, there is tension between the relationship in which the model represents the actual problem and the simplifications required to keep the processes treatable. However, the inherent simplification of the entire model provides a number of advantages. Simplification allows, for example, decision-makers to improve their understanding of problems by concentrating on key issues (Cairns et al., 2016).

Simple understanding processes in modeling may be more attractive to decision-makers who tend to reject mathematical methods or do not want to employ the time and effort required by more complex formulations (Cairns et al., 2016). Thus, the use of an approximate but simple model with recognized limitations may be preferable to decision-makers (Keeney, 2004).

In addition, in the modeling process there is a tendency for the modeler to insert a large number of variables, in order to make the model devised more representative and complete. This practice is not advisable, the inclusion of only the most significant factors being recommended in order to make it less complex. Instead of trying to create, in the initial stage, a complex model that incorporates all aspects of the situation analyzed, it is best to start with a simple model. In this way it is possible to provide learning, by which a simple model can be gradually refined, if necessary (Pidd, 1996).

The main advantages of using a simplified model are: (i) the time savings achieved in execution and planning; (ii) the logic of the model being more easily understood by the people involved; and (iii) the ability of the model to be reformulated quickly, if necessary. An important lesson in constructing models for decision support is that they serve as learning tools. It is recommended that the organization should be made by the model constructor but following the requirements and advice of the system user or the decision-maker, in general, the problem owner (Wierzbicki, 2007).

6.2   DEA MODELING METHOD

The process of designing each model is the subject of much discussion by researchers and professionals in general. In the design of models for analysis using DEA, this point applies in particular (Cook et al., 2014; Liu et al., 2016). There has been research contributing to the modeling stage in DEA. For example, Jain et al. (2011) proposed a framework to define variables and the DEA model, according to Figure 6.1.

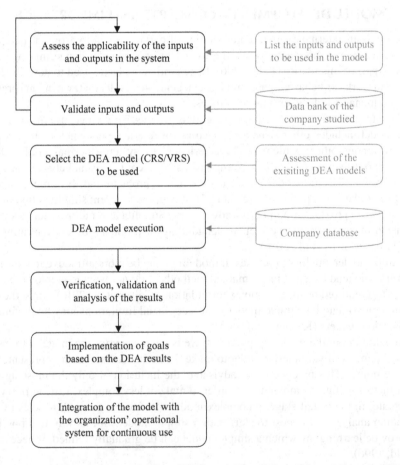

FIGURE 6.1   Framework for selection of DEA variables. (Adapted from Jain et al., 2011.)

For Jain et al. (2011) there are two main steps necessary for the execution of DEA. First, an adequate set of inputs and outputs needs to be determined and used, according to the appropriate DEA model. The two steps described require involvement and discussion between specialists and business professionals. In addition, it is recommended, as far as possible, to limit the number of variables to be considered in the model, as this allows for better discrimination, facilitating the identification of efficient and inefficient decision-making units (DMUs).

In Figure 6.1, it is shown that the process of defining variables of the DEA model should start with a list of the possible inputs and outputs of the process to be analyzed. Subsequently, it should be determined which of these related variables are more closely aligned with the process under study. With this, the availability of information in the company in which it is intended to develop the study is determined, and, afterward, analysis and validation of the data can be carried out. It is common that the planned conceptual model cannot be applied in full due to the unavailability of data on the organization. Nevertheless, it provides a good orientation for the initial process of definition of variables.

However, it should be noted that the framework proposed by Jain et al. (2011) is limited to helping researchers with the definition of variables and the DEA model to be used in the analysis. Other steps required for DEA modeling (definition of the units of analysis, period of analysis, DMU and model orientation, for example) are not addressed by Jain et al. (2011). Thus, it is noted that the framework proposed by the authors is limited to a part of the procedures necessary for the application of DEA. This restricts its application by researchers and professionals. In addition, the framework does not provide a holistic systemic view of the DEA modeling process.

We propose a method that takes into account all DEA modeling phases. We call the method MMDEA (Modeling Method in DEA). The MMDEA is summarized in Figure 6.2. It is understood that the MMDEA is a contribution to the theory of the application of DEA, since the definition of the model is one of the main challenges in the use of the technique. In addition, the proposed method considers interaction between researchers and companies in their application. In MMDEA, we suggest execution of these steps involving researchers, one or more focus groups and managers of the companies in which the analysis will be carried out. This interaction between the researchers and the company is considered to be important because it brings theory and practice closer together. MMDEA was developed using ARIS (Scheer, 1994).

MMDEA consists of steps and sub-steps outlined in the proposed method, which will be discussed in the following sections.

6.2.1 Define the Purpose of the Analysis and the Type of Efficiency to Be Assessed

Step 1 of the MMDEA includes identification of the problem to be analyzed and the definition of the purpose of the analysis. The application of DEA and consequently the modeling requires applications to concrete cases. Thus, identification of the problem can also be considered to be the definition of the context in which the analysis of productivity and efficiency will be carried out. To better understand the problem, the Restriction Theory Thought Process (TOC) and Systemic Thought can be used.

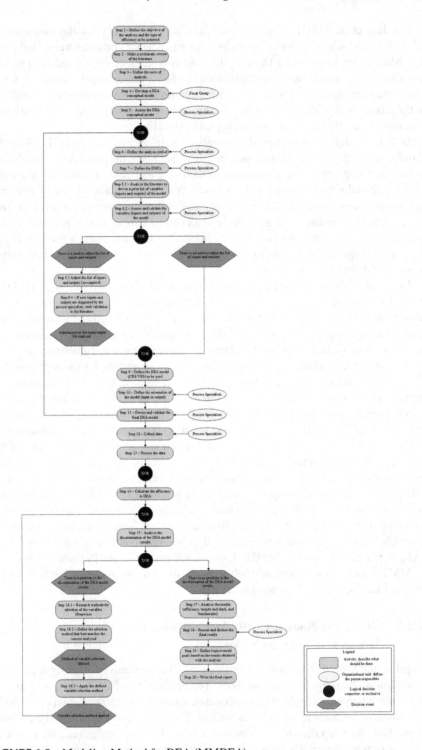

FIGURE 6.2 Modeling Method for DEA (MMDEA).

In addition, it is necessary to define which type of efficiency will be assessed, namely technical efficiency or its variations, namely scale, allocative or cost/economic efficiency. It is important for this definition to be made as early as Step 1, as it will guide the systematic review of the literature in Step 2. Preliminary planning is also recommended, taking into account the resources and the length of time of the project design.

6.2.2 Carry Out a Systematic Literature Review

Step 2 consists of a systematic review of the literature (RSL). The DEA application process can be facilitated by the identification in the literature of analyses similar to the planned one. Thus, we recommend a systematic review of the literature on the topic of interest, including books, articles, dissertations, theses and other research material found in national and international databases.

In the systematic literature review, one may find, for example, a similar analysis that can support the definition of the DMUs or that provides a list of inputs and outputs to be used in the planned analysis. To review the literature of Step 2, we recommend the procedures devised by Morandi and Camargo (2015).

6.2.3 Define the Units of Analysis

In Step 3, it is necessary to define the units of analysis, both for internal and external benchmarking. The unit of analysis is the element to be analyzed, in its particular context. Care should be taken not to confuse the unit of analysis with the DMU. For instance, in a longitudinal internal benchmark (over time), the unit of analysis may be the engineering of a company's products, and the DMU the monthly batch of projects developed by the Product Engineering department. The unit of analysis is defined in a more general context and the DMU in a more specific context. The definition of the units of analysis can be made with the support of the specialists of the company studied.

As for internal benchmarks, we caution that the assessments of the operations may not reflect the theoretical performance, even for those DMUs considered to be efficient. This is because, in DEA, one or more DMU benchmarks are identified within the set of DMUs analyzed. For example, if an assessment considers a department over time (e.g. two years), the DMU considered to be efficient will be the one with the best performance over this period.

6.2.4 Developing a DEA Conceptual Model

In Step 4, we suggest the development of a conceptual model of the planned analysis. A conceptual model is a simplified representation of a real system or context, which considers variables and concepts and the relationship between them. In the specific case of the development of the DEA conceptual model, it allows a better understanding of the context of the analysis. The conceptual model can be represented by a research design that helps with the understanding and illustration of the analysis that will be carried out. In Chapter 7, where we demonstrate how to apply MMDEA, we will provide an example that illustrates a research design.

The main objective of the conceptual model, with the elaboration of the research design, is to facilitate communication between the researcher/modeler and the other interested parties in the situation being analyzed. For definition of the conceptual model, it is suggested that the researcher/modeler seeks support from a focus group. For composition of the focus group, it is recommended that at least one participant has knowledge about the DEA technique and another knows in advance the processes where the analysis will be performed. This step is important because it can help the researcher/modeler to prepare an initial presentation for a group of specialists from the company or companies for which the analysis is intended. In this way, it is possible to discuss in greater depth the conceptual research model with the company's specialists.

6.2.5 ASSESSMENT OF THE DEA CONCEPTUAL MODEL

Step 5 consists of presenting the conceptual framework of research to a group of specialists in the company where the analysis will be developed. The purpose of this initial discussion is to assess the conceptual model which has been developed. This is important at this stage, because, if the conceptual model is poorly developed, the whole analysis will probably be prejudiced. Thus, it is understood that the validation of a model, mainly involving managers and specialists, contributes to reduction of distortions in the research process and increases the credibility of the conclusive analysis of a study (Jonsen & Jehn, 2009; Sodhi & Tang, 2014). The DEA literature itself recommends the use of specialists in the process under analysis (Jain et al., 2011; Park et al., 2014; Piran et al., 2016).

It is recommended that specialists selected to participate in the study be chosen based on their experience and knowledge of the company's processes and for their ability to support the development of the project, as well as in assisting with the data collection. The team formed to participate in the project should be multidisciplinary, since the professionals involved should hold strategic positions (for example, directors), tactical positions (for example, managers) and operational positions (for example, engineers). This multi-disciplinarity is important because different organizational positions influence individuals differently in their interpretation of events (O'Leary-Kelly & Vokurka, 1998).

If it is not possible to use the focus group, it can be deleted from the method. However, the involvement of process specialists is strongly recommended. This is because, in a modeling process, there is no way to know if all the variables considered significant were included in the model devised. Thus, it is necessary to make sure that the variables considered more representative by the decision-making group are present. We also suggest that, if the process specialists do not know the DEA concepts, a conceptual leveling is made by the researcher/modeler.

6.2.6 DEFINITION OF THE TIME PERIOD OF THE ANALYSIS

Step 6 consists of defining the time period of the analysis, which depends on whether the benchmark is internal or external. If internal, one should make an analysis over time, called a longitudinal analysis. In such a case, the DMU will be compared to itself over time. In addition, an analysis over time is also necessary when aiming to observe the effect of an intervention in the unit of analysis, or when there is a delay in relation to the effects of interventions made in the unit of analysis. Therefore, the definition of the period of analysis is relevant and should be as long as possible. Normally, days, months or years are used as base periods.

In the external benchmarks, cross-sectional analyses, i.e. of several different DMUs, are usually performed over the same period of time. For example, if we perform an analysis among the automakers of Brazil, we can use the current period as a base, and the variation will be in relation to the different existing manufacturers (Ford, Toyota, Fiat, GM, etc.). Besides this, it is possible to perform a longitudinal and transversal analysis together, that is, over time and with different DMUs. These cases are called analyses with panel data. For example, one can perform an analysis of the vehicle assemblers (Ford, Toyota, Fiat, GM, etc.) over the past five years of each automaker's operations.

6.2.7 DEFINING THE DMUs

Step 7 consists of the definition of the DMUs. Putting it simply, the DMUs represent what will be compared. If there is difficulty identifying the DMUs, the question should be asked: What do I want to compare?

As we have shown previously, the DMU is different from the unit of analysis. For internal benchmarks, the DMU can be the monthly batch of products, among others. For external benchmarks, the DMUs represent the companies or departments of different organizations about which the comparison will be made. It is important to note that, for the execution of the analysis using DEA, which is based on a benchmark, the set of DMUs adopted should consider the same variables of inputs and outputs, varying only in quantities used. Furthermore, it should be homogeneous, that is, performing the same tasks, with the same objectives. In summary, the DMUs may have distinct scales, but should be comparable. It is not possible to obtain valid results when comparing a vehicle assembler with a supermarket chain, for example. The DMU can be defined with support from the existing literature and with the support of the specialists of the company in which the analysis is being developed.

6.2.8 DEFINING THE VARIABLES TO BE USED IN THE DEA MODEL

Step 8 consists of defining the variables to be used in the DEA model. The definition of the input and output variables can be considered to be the most important step in the modeling process using DEA (Wagner & Shimshak, 2007). Cook et al. (2014) emphasized that, in analyses using DEA, it is not possible to have complete certainty that all the significant variables were included in the model described. However, every effort should be made to include variables that make practical sense for the proposed research.

Dyson et al. (2001) argued that, in defining a set of input and output variables, one should seek to respect fundamental assumptions, such as: (i) to cover the widest possible range of resources used in the analytical process under review, but to avoid making the model too complex; ii) to capture all levels of activities and the maximum possible performance measurements; iii) to define a set of variables common to all units of analysis; and iv) to consider environmental variables, if necessary.

Considering that the definition of variables is acknowledged to be a central issue in DEA modeling, researchers may tend to define a large number of variables to ensure that the DEA model developed adequately represents the process analyzed. However, if a DEA model has a relatively large number of variables in relation to the number of defined DMUs, it will tend to present a problem of discrimination among DMUs (Adler & Yazhemsky, 2010). Thus, it is necessary to construct a DEA model with the variables considered to be significant in the proposed analysis (Wagner & Shimshak, 2007).

The literature on criticality in defining the variables of DEA models is vast (e.g. Dyson et al., 2001; Adler & Yazhemsky, 2010; Cook et al., 2014). Seeking to contribute in this way, our method proposes definition of variables by the following sub-steps:

i)   Analyze the literature to determine a prior listing of the variables (inputs and outputs) of the model.
ii) Assess and validate the variables (inputs and outputs) of the model with process specialists.
iii) If necessary, make adjustments to the variables listed.
iv) If new variables are suggested by the process specialists, seek validation in the literature beforehand.

As can be seen, a circular process, integrating theory and practice, is defined. First, our recommendation is for the researcher/modeler to begin the discussion with process specialists with a prior listing of variables that can be used. This listing can be developed based on the selected studies in the review of the literature carried out in Step 2. If necessary, other literature reviews can be made with a focus on the detection of variables. This pre-listing is often very useful, as frequently the company's specialists have many questions at the beginning of the modeling process. Thus, this prior listing of variables helps and speeds up this definition. Additionally, it becomes possible to pre-validate model variables with the specialists.

Another point to note is that, in many cases, there are no data available for the previously listed variables. This lack of data will be pointed out by the company's specialists at the time of the discussion of the variables. When data are unavailable, it is necessary to delete the corresponding variable from the model. Another aspect is that the specialists may suggest other variables that they deem important for the context analyzed and which were not included in the previous list of variables determined by the researcher/modeler. In these cases, it is important to keep the traceability of the variables inserted in or excluded from the process, keeping a record of the reasons that justify these actions.

These recommendations are very important because the specialists usually know the analysis context better than does the researcher/modeler. However, if new variables are suggested by the company specialists, we recommend that the researcher/modeler seek the validation of these variables in the literature before incorporating them into the model. The researcher/modeler should also observe important aspects, such as making the results of the process of defining variables comparable with national and international studies.

6.2.9 DEFINING THE DEA (CRS/VRS) MODEL TO BE USED

Step 9 consists of defining the DEA model to be used. This definition should be made by the researcher/modeler based on the concepts of the DEA (CRS and VRS) models. In order to make such a decision, it is necessary to seek information about the context analyzed to understand the scale proportionality relations between inputs and outputs and the DMUs analyzed.

6.2.10 DEFINE THE ORIENTATION OF THE MODEL (INPUT OR OUTPUT)

Step 10 consists of defining the orientation of the model. As mentioned earlier (Chapter 5) there are two possibilities for the orientation of the DEA model: input (input) or output. It is to be remembered that the orientation to input occurs when it is sought to minimize the use of resources (input variables) and keep the outputs constant. In this situation, the DEA results returned to the researcher and decision-maker indicate the amount of input resources established in the model that could be used and saved by increased efficiency. Output orientation occurs when one tries to keep the inputs constant and maximize the outputs. In this case, the DEA results returned to the researcher and decision-maker indicate the amount that could be produced consuming the same amount of resources (Cook et al., 2014).

The definition of the orientation of the model depends entirely on the context and purpose of the analysis and should be defined by the process specialists and the company in which the analysis is being developed. If necessary, the researcher/modeler can also participate in this decision. For example, if the company needs to increase its capacity to generate products, projects or services, since it has market demand for this, it is possible to recommend orientation to output. In this case, the analysis will calculate the efficiency and will seek to show through the targets and slack how much more the company can produce using the same resources.

If the company needs to reduce the consumption of resources to continue the same level of production, the suggested orientation is the input. Input orientation is recommended when inputs are more controllable than outputs, which may depend on the market, for example (Hamdan & Rogers, 2008). In situations of low market demand, input-oriented analyses may also be of interest.

6.2.11 ELABORATION AND VALIDATION OF THE FINAL DEA MODEL

Step 11 consists of the elaboration and validation of the final DEA model. After completing the above steps, we suggest that the researcher/modeler draws an illustration of the DEA model defined and makes a presentation to the process specialists. A suggested illustration is shown in Figure 6.3.

The illustration presented in Figure 6.3 shows all the decisions made in the modeling process organized in a simple way to aid understanding. Initially, the unit of analysis can be described. Subsequently, a description of the inputs used can be

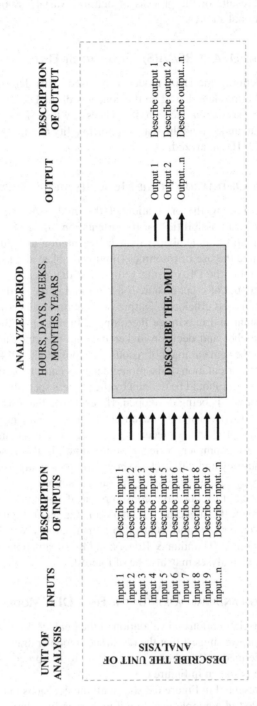

FIGURE 6.3   Illustration of a DEA model.

made, simulating a system in which these inputs are inserted in the DMU, where they are processed and transformed into the output(s), i.e. the products/services resulting from the process. This schematic illustration helps the whole process of elaboration and presentation of the analysis.

6.2.12 COLLECTING DATA

Step 12 consists of the data collection process. When testing or constructing theories from case studies, researchers should have a clear focus on collecting specific data in a systematic way (Mintzberg, 1979). Although research questions evolve throughout the work, focus and coherence should be maintained throughout the data collection and analysis (Eisenhardt, 1989; Voss et al., 2002). Usually, the data are collected with support from the specialists of the company studied. We underline the importance of using information extracted from the companies' ERPs. In addition, data is typically collected from information recorded on spreadsheets. The important point is to give preference to observed data over perceptual data.

6.2.13 HANDLING THE DATA

Step 13 consists of the handling of the data collected. Initially, the information can be organized on spreadsheets. These will be used to insert the data into the software we developed and will be shown in Chapter 8. We have provided an example in Chart 6.1 of how information can be organized.

Initially, a column is created with the numbering of the DMUs defined. DMUs can be coded to facilitate analysis and interpretation of results, as well as to avoid exposing confidential company information. Subsequently, rows with inputs and outputs should be created. Thus, it is possible to fill in the rows with the information regarding the quantities of each input and output of each DMU.

CHART 6.1
Example of Spreadsheet for Data Organization

DMU	INPUT1	INPUT2	INPUT3	INPUT4	INPUT5	INPUT6	OUTPUT1
1	26	58	20	12,080	4,100	7,980	2
2	163	243	134	68,780	24,352	44,428	11
3	242	422	221	108,928	38,568	70,360	18
4	147	250	144	85,170	29,148	56,002	14
5	165	300	130	91,900	32,160	59,740	15
6	74	75	70	49,460	18,099	31,361	8
7	134	142	121	74,234	25,534	48,700	12
8	119	126	125	54,362	18,178	36,184	9
9	78	123	110	42,644	14,364	28,280	7
10	118	215	189	79,844	27,256	52,588	13

6.2.14　CARRYING OUT THE EFFICIENCY CALCULATION IN DEA

Step 14 consists of the calculation of efficiency in DEA. To carry out such a calculation, we suggest the SAGEPE software (System for Analysis and Management of Productivity and Efficiency) that was developed by the authors of this book. In Chapter 8, we will show, step by step, how to use this software.

6.2.15　ANALYZING THE BREAKDOWN OF THE DEA MODEL RESULTS

Step 15 consists of analysis of the discrimination of the results of the DEA model developed. When it is not possible to differentiate efficient DMUs from the inefficient, it is concluded that the model presents a problem of discrimination. In other words, the discrimination problem is observed when all the DMUs have an efficiency equal to or very close to 1 (100%). When the situation is opposite to that described, it is concluded that the model does not present a discrimination problem. For a better understanding, see Chart 6.2 with an analysis of each case.

When a model presents a discrimination problem, it is necessary to use a method for the selection of variables, also known as Stepwise. When the model does not present discrimination problems, it is possible to go straight to the analysis of the results (Step 17).

6.2.16　USING A SELECTION METHOD FOR VARIABLES (STEPWISE)

Step 16 is the use of a Stepwise selection method. With the application of Stepwise, there is refinement of the DEA model. Thus, this objective aims, through literature research, to determine the use of a method that helps the researcher define whether there are variables that can be considered insignificant, and can be excluded from the model, in order to avoid problems of discrimination efficiency scores. For this phase, three fundamental sub-steps need to be fulfilled:

i) Search for methods of variable selection (Stepwise) in the literature.
ii) Define the selection method for variables that best fits the context analyzed.
iii) Apply the defined variable selection method.

CHART 6.2
Example of Discrimination Analysis of Efficiency Scores

DMUs	Efficiency WITH a discrimination problem	Efficiency WITHOUT a discrimination problem
1	1.00	1.00
2	1.00	0.9931
3	1.00	0.8414
4	1.00	1.00
5	0.9900	0.7888
6	0.9929	0.8162

These steps are important because the selection of the ideal method really depends on the context of analysis. The DEA modelling does not provide precise guidance for specifying input and output variables. Instead, such definitions are left to the discretion of the user (Nataraja & Johnson, 2011). However, even if it is understood that the DEA model loses its discriminatory power with a high number of variables (inputs and outputs added), we suggest not limiting the participation by and suggestions of the focus group and company specialists in the process of defining inputs and outputs which are considered to be significant for inclusion in the analysis.

When assessing the omission or inclusion of variables considered important, it is understood to be preferable to add variables considered significant at the time of the DEA modeling and, later, to apply selection methods for variables (Adler & Yazhemsky, 2010). Some of the methods for selecting identified variables are summarized in Chart 6.3.

The verification of the methods presented in Chart 6.3 indicates the use of statistical procedures, considering analyses for the removal or inclusion of variables or tests of correlation among variables. Although not a general rule, we suggest that readers use two methods. First, we suggest use of the efficiency contribution measure (ECM), developed by Pastor et al. (2002).

ECM was designed to analyze the significance of each variable used in the model based on its contribution to efficiency. The variable to be tested is called the candidate variable. Thus, the ECM approach is based on two variable specifications, called total model and reduced model. The total model contains all the variables defined in the model. In the reduced model, one of the input or output variables (the candidate variable, also called "z") should be removed.

With the removal of the candidate variable, the DMU efficiency score, ECM, of the variable "z", denoted by φ, represents the change in efficiency scores between the total model and the reduced model (Eskelinen, 2017). If $\varphi = 1$, we can conclude that the candidate variable "z" has no influence on the efficiency assessment of the DMU. Similarly, $\varphi = 1.1$ indicates that removal of the candidate variable has a 10% impact on the DMU efficiency score. Such scores are calculated for all the DMUs (Eskelinen, 2017).

CHART 6.3
Methods for Selection of Variables in DEA

Method	Author(s)
PCA DEA - Principal component analysis	Ueda and Hoshiai (1997); Adler and Golany (2001)
Bootstrapping for Variable Selection	Simar and Wilson (2001)
ECM - Efficiency contribution measure	Pastor et al. (2002)
Recursive method	Fanchon (2003)
A regression-based test	Ruggiero (2005)
Selecting variables based on partial covariance	Jenkins and Anderson (2003)
Progressive or "Stepwise" Selection Process	Wagner and Shimshak (2007)

The relevance of the candidate variable "z" for the total model can be tested statistically using a binomial test. The null hypothesis is that the efficiency scoring distributions of the original model and the reduced model do not differ significantly and that the candidate variable could be eliminated from the model.

The test has two parameters. The first is a parameter for tolerated changes between the models (called φ^-). So, T is the number of DMUs that exceed this tolerance. The second parameter is the probability level (denoted by p0) for the maximum proportion of DMUs allowed that exceed the tolerance. For the ECM, a candidate variable is considered significant for the process under analysis if the percentage value of p0 in relation to the processes analyzed achieve a change of efficiency that is greater than φ^-. Therefore, p0 = 15% and $\varphi^- = 10\%$ are selected, as suggested by Pastor et al. (2002). These parameters mean that the proportion of DMUs more than 10% affected should not exceed 15% at the level of significance established. Otherwise, the variable cannot be eliminated (Eskelinen, 2017).

Finally, two approaches are defined: i) direct selection (addition of variables); and ii) direct elimination (removal of variables). We suggest using the direct elimination process, with which the analysis starts, presenting all the variables determined for the model, and the removal of each variable is performed and analyzed. We emphasize that all the procedures described should be conducted using the software SAGEPE that we have developed. In this way, these procedures can be performed easily.

Secondly, we suggest the method developed by Wagner and Shimshak (2007). These authors have developed a method for a process of progressive selection of variables, which they call Stepwise. In the Stepwise method, the variables removed or added to the model are considered the mean efficiency variation. This method is intended to aid refinement of DEA models, so that only variables that impact the efficiency of the model in use are considered.

Stepwise starts by considering all the input and output variables listed in the DEA model. Subsequently, at each step, a variable is removed from the model by means of analysis of the efficiency scores of the DMUs. The steps suggested by the authors (Wagner & Shimshak, 2007) are as follows:

Step 1: Execute the DEA analysis on the model originally defined, that is, considering all inputs and outputs (represented by E * 0).

Step 2: Record the efficiency scores of each DMU for the analysis performed.

Step 3: Perform the calculation of the arithmetic mean of the scores of the DMU efficiencies analyzed (represented by Ex * 0).

Step 4: Perform the DEA analysis by removing one variable at a time from the model (for example, remove variable 1 and perform DEA analysis, then put variable 1 back into the model and remove variable 2 by running the DEA analysis again). Repeat the process successively until all variables have been considered (represented by E * 1, E * 2, ..., En).

Step 5: Record the efficiency scores of each DMU for each analysis performed, considering the withdrawal of each variable.

Step 6: Calculate the arithmetic mean of the DMU efficiency scores analyzed, in which the withdrawal of one variable is considered (represented by Ex * 1, Ex * 2, ..., Exn).

Step 7: Calculate the difference between the mean efficiency resulting from the analysis considering the original model (with all the variables) and the analyses performed in Step 3 (taking one variable at a time), i.e. Ex*0 - Ex*1, Ex*0 - Ex*2, ..., Ex*0 - Ex*n.

After performing the calculations, it is necessary to verify which analysis/es (Ex*0 - Ex*1, Ex*0 - Ex*2, ..., Ex*0 - Ex*n) describe(s) the least variation among the means of the efficiencies. In this sense, it is understood that the variables that presented less variation, or where variation was equal to 0, are not contributing significantly to the efficiency and can be excluded from the model. After excluding variables that do not impact the model, the researcher should return to Step 1 of the procedure and restart Stepwise. This Stepwise analysis, in which variables can be excluded, can be performed until the model presents a single input and a single output (Wagner & Shimshak, 2007).

After identifying the important variables to be used in the DEA model, we suggest that the researchers/modelers make a final assessment. Another important point is to carry out the adjustments in the model previously illustrated in Figure 6.3. This is because the variables used in the model previously made in the "Elaborate and validate the final DEA model" step may change after using Stepwise and the assessment of the specialists. Thus, the illustration of the model should be updated, and the modeling process can then be considered to be complete.

6.2.17 ANALYZE THE RESULTS (EFFICIENCY, TARGETS AND SLACK AND BENCHMARKS)

Step 17 consists of analyzing the results obtained. One of the major challenges of data analysis is that the researcher/modeler should objectively demonstrate the process by which data and field notes were developed into conclusions (Barratt et al., 2011). We suggest that graphical analyses be used to facilitate understanding and illustration of the efficiency scores. Also, we suggest analyses of the targets and slack and benchmarks, which can provide valuable information for the researchers and managers. In addition, complementary techniques are commonly used to aid understanding of the results obtained through DEA. We will show some of these techniques in Chapter 9.

6.2.18 PRESENT AND DISCUSS FINAL RESULTS

Step 18 consists of presenting and discussing the results. With the results obtained and analyses performed, we understand that it is important to conduct a final discussion with the process specialists who supported the development of the study by participating in and assisting with the entire process of modeling and data collection. The objective of this discussion is to demonstrate the results found in the analysis for these specialists. In addition, presentation of the specialists' knowledge regarding the process analyzed will provide contributions to give meaning to the analyses made by the researcher/modeler. This process is very important and helps the researcher/modeler to understand more clearly the figures related to the efficiency scores obtained and to write a final report of higher quality.

6.2.19 DEFINING IMPROVEMENT GOALS, BASED ON THE RESULTS OBTAINED FROM THE ANALYSIS

Step 19 consists of definition of the improvement goals based on the results obtained in the analysis. At this stage, the researcher/modeler, together with the managers of the organization under study, should draw up action plans to seek the improvements indicated by the analysis. The analysis of targets and slack provides valuable assistance in devising these plans.

6.2.20 WRITING THE FINAL REPORT

Finally, in Step 20, the researcher/modeler should concentrate on writing the final report to communicate the findings of the analysis. In the report, it is important to make clear all the procedures performed in relation to the method used in the development of the work. Moreover, it is necessary to objectively demonstrate the results and how they can contribute to improvement of the management of the organization. The final process of drafting the report should not be underestimated, since a lack of care in reporting the results may negatively affect all the work carried out.

REFERENCES

Adler, N., & Golany, B. (2001). Evaluation of deregulated airline networks using data envelopment analysis combined with principal component analysis with an application to Western Europe. *European Journal of Operational Research*, *132*(2), 260–273.

Adler, N., & Yazhemsky, E. (2010). Improving discrimination in data envelopment analysis: PCA–DEA or variable reduction. *European Journal of Operational Research*, *202*(1), 273–284.

Barratt, M., Choi, T. Y., & Li, M. (2011). Qualitative case studies in operations management: Trends, research outcomes, and future research implications. *Journal of Operations Management*, *29*(4), 329–342.

Cairns, G., Goodwin, P., & Wright, G. (2016). A decision-analysis-based *framework* for analysing stakeholder behaviour in scenario planning. *European Journal of Operational Research*, *249*(3), 1050–1062.

Cook, W. D., Tone, K., & Zhu, J. (2014). Data envelopment analysis: Prior to choosing a model. *Omega*, *44*, 1–4.

Dyson, R. G., Allen, R., Camanho, A. S., Podinovski, V. V., Sarrico, C. S., & Shale, E. A. (2001). Pitfalls and protocols in DEA. *European Journal of Operational Research*, *132*(2), 245–259.

Eisenhardt, K. M. (1989). Building theories from case study research. *Academy of Management Review*, *14*(4), 532–550.

Eskelinen, J. (2017). Comparison of variable selection techniques for data envelopment analysis in a retail bank. *European Journal of Operational Research*, *259*(2), 778–788.

Fanchon, P. (2003). Variable selection for dynamic measures of efficiency in the computer industry. *International Advances in Economic Research*, *9*(3), 175–188.

Hamdan, A., & Rogers, K. J. (2008). Evaluating the efficiency of 3PL logistics operations. *International Journal of Production Economics*, *113*(1), 235–244.

Jain, S., Triantis, K. P., & Liu, S. (2011). Manufacturing performance measurement and target setting: A data envelopment analysis approach. *European Journal of Operational Research*, *214*(3), 616–626.

Jenkins, L., & Anderson, M. (2003). A multivariate statistical approach to reducing the number of variables in data envelopment analysis. *European Journal of Operational Research, 147*(1), 51–61.

Jonsen, K., & Jehn, K. A. (2009). Using triangulation to validate themes in qualitative studies. *Qualitative Research in Organizations and Management: An International Journal, 4*(2), 123–150.

Keeney, R. L. (2004). Making better decision makers. *Decision Analysis, 1*(4), 193–204.

Liu, J. S., Lu, L. Y., & Lu, W. M. (2016). Research fronts in data envelopment analysis. *Omega, 58*, 33–45.

Mintzberg, H. (1979). An emerging strategy of "direct" research. *Administrative Science Quarterly, 24*(4), 582–589.

Morandi, M. I. W. M, & Camargo, L. F. R (2015). Systematic literature review. In *Design science research: A method for science and technology advancement* (pp. 129–158). New York: Springer.

Nataraja, N. R., & Johnson, A. L. (2011). Guidelines for using variable selection techniques in data envelopment analysis. *European Journal of Operational Research, 215*(3), 662–669.

O'Leary-Kelly, S. W., & Vokurka, R. J. (1998). The empirical assessment of construct validity. *Journal of Operations Management, 16*(4), 387–405.

Park, J., Lee, D., & Zhu, J. (2014). An integrated approach for ship block manufacturing process performance evaluation: Case from a Korean shipbuilding company. *International Journal of Production Economics, 156*(1) 214–222.

Pastor, J. T., Ruiz, J. L., & Sirvent, I. (2002). A statistical test for nested radial DEA models. *Operations Research, 50*(4), 728–735.

Pidd, M. (1996). *Modelagem empresarial: ferramentas para tomada de decisão.* Bookman.

Piran, F. A. S., Lacerda, D. P., Camargo, L. F. R., Viero, C. F., Dresch, A., & Cauchick-Miguel, P. A. (2016). Product modularization and effects on efficiency: An analysis of a bus manufacturer using data envelopment analysis (DEA). *International Journal of Production Economics, 182*, 1–13.

Ruggiero, J. (2005). Impact assessment of input omission on DEA. *International Journal of Information Technology & Decision Making, 4*(03), 359–368.

Scheer, A. W. (1994). Architecture of integrated information systems (ARIS). In *Business process engineering* (pp. 4–16). Berlin, Heidelberg: Springer.

Simar, L., & Wilson, P. W. (2001). Testing restrictions in nonparametric efficiency models. *Communications in Statistics-Simulation and Computation, 30*(1), 159–184.

Sodhi, M. S., & Tang, C. S. (2014). Guiding the next generation of doctoral students in operations management. *International Journal of Production Economics, 150*, 28–36.

Sterman, J., Oliva, R., Linderman, K., & Bendoly, E. (2015). System dynamics perspectives and modeling opportunities for research in operations management. *Journal of Operations Management, 3940*, 1–5.

Ueda, T., & Hoshiai, Y. (1997). Application of principal component analysis for parsimonious summarization of DEA inputs and/or outputs. *Journal of the Operations Research Society of Japan, 40*(4), 466–478.

Voss, C., Tsikriktsis, N., & Frohlich, M. (2002). Case research in operations management. *International Journal of Operations & Production Management, 22*(2), 195–219.

Wagner, J. M., & Shimshak, D. G. (2007). Stepwise selection of variables in data envelopment analysis: Procedures and managerial perspectives. *European Journal of Operational Research, 180*(1), 57–67.

Wierzbicki, A. P. (2007). Modelling as a way of organising knowledge. *European Journal of Operational Research, 176*(1), 610–635.

7 Illustrating the Application of MMDEA

This chapter presents two practical applications of the MMDEA. One considers a research program carried out in a goods production system (Section 7.1). The other takes account of research carried out in a service provision system (Section 7.2).

7.1 APPLICATION OF THE MMDEA IN A SYSTEM OF PRODUCTION OF GOODS

Initially, we will show the application of the proposed method in an analysis performed on a goods production system. The application presented refers to the practical problem of efficiency analysis described in Chapter 3, Section 3.1. The focus will be on the application of the modeling method, and the results of the research will be shown in Chapter 10, Section 10.1

7.1.1 DEFINING THE PURPOSE OF THE ANALYSIS AND THE TYPE OF EFFICIENCY TO BE ASSESSED

The research was conducted at a bus manufacturing company that had been implementing product modularization since 2007. Product modularization (whose concepts are presented in Chapter 11) is usually described as a general set of principles for managing complexity in product processes. The complexity of products and processes has contributed, for example, to reduced productivity and efficiency in manufacturing systems. Considering that modularization assists in the management of complexity, a central aspect in the research on modularization is to assess the effects of its use on the productivity and efficiency of organizations. The research carried out sought to understand whether the implementation of product modularization really helped in managing the complexity of products and processes, and achieved improvements in productivity and efficiency of the company for which the analysis was developed. Thus, in the investigation carried out, the focus of the analysis had the following general objective:

- To analyze the effect of product modularization on the efficiency of the Product Engineering and Production Process of a bus manufacturer.

The specific objectives of the work were as follows:

- To assess the efficiency behavior of the Product Engineering and Production Process over time, comparing the period before and after the implementation of modularization.
- To assess whether there was a significant difference between the mean of the Product Engineering and Production Process efficiencies, comparing the periods before and after the modularization.
- To quantify the effects of modularization observed on the efficiency of the Product Engineering and the Production Process over time.
- To establish the causality between modularization and effects on the efficiency of the Product Engineering and the Production Process.

It was defined that the type of efficiency to be analyzed was the composite technical efficiency.

7.1.2 EXECUTION OF A SYSTEMATIC REVIEW OF THE LITERATURE

According to orientation of the MMDEA, a systematic review of the literature was carried out and a research design was elaborated. The literature review aimed to identify studies that assessed the overall effects of modularization on productivity and efficiency, in particular. In order to carry out this research, the following keywords were used: modularization, modularity, modular, modular product, modular production, modular project and modular service. These keywords were combined with data envelopment analysis, productivity, efficiency, and performance. The research was carried out with these combinations of words in both Portuguese and English.

7.1.3 DEFINITION OF THE UNITS OF ANALYSIS

Two units of analysis were defined to carry out the research. The first was Product Engineering and the second, the Production Process of the company. From these definitions, two DEA models were developed, one of them considering the characteristics of Product Engineering and the other taking into account the characteristics of the Production Process.

7.1.4 DEVELOPMENT OF A DEA CONCEPTUAL MODEL

From the systematic review of the literature, it was possible to identify research that aided development of the conceptual DEA models. The conceptual models were developed with the support of a focus group. The focus group contained one specialist in the application of the DEA technique and a consultant specialized in the processes of the bus manufacturing company. This group provided preliminary guidance regarding definition of the variables and the process of data collection and processing. Thus, two research designs were developed, one for Product Engineering (Figure 7.1) and another for the Production Process (Figure 7.2).

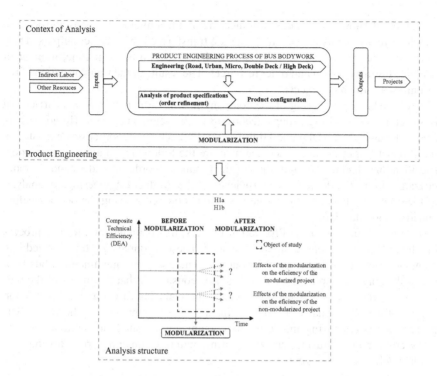

FIGURE 7.1   Research design of the Product Engineering.

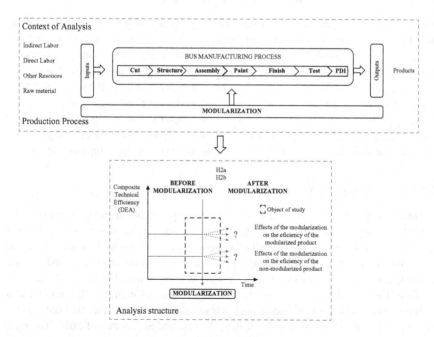

FIGURE 7.2   Research design of the Production Process.

Figure 7.1 shows the Product Engineering of a bus manufacturer as a process in which there are inputs of resources that are transformed into project outputs. Thus, inputs can be considered to be resources that take into account time, number of people, plus information and other resources. The results of the transformation of the resources are the bus projects.

The Product Engineering of the company in which the research was carried out is structured in working groups according to the segments served (Road, Urban, Micro, Double Deck (DD) and High Deck (HD)). The macroprocesses executed in the Product Engineering of each segment are the following: (i) analysis of product specifications; and (ii) configuration of the product. In order to understand if modularization impacted the technical efficiency of the Product Engineering, an analysis of the efficiency behavior over time was carried out, comparing the periods before and after modularization.

As for the analysis of the Production Process, this is understood to be a process in which there are resource inputs ("inputs"). These resources are transformed into manufacturing, resulting in product outputs. The company manufactures bus bodies on the chassis supplied by customers. It is considered that the inputs of the production process are resources, such as raw materials and direct and indirect labor, among others. The final products ("outputs") of the company are the buses. The process of manufacturing and production organization is subdivided into the following sectors: manufacturing, cocoon, plating, painting, finishing, road-testing, and product delivery inspection (PDI).

7.1.5   Assessment of the DEA Conceptual Model

The conceptual model developed was assessed by a group of professionals designated by the organization to support the research. These professionals are listed in Chart 7.1. This chart also summarizes the support function exercised by each company specialist and the service time of each professional in the organization.

These professionals were chosen based on their experience in the segment, knowledge of the company processes, ability to support the development of the project, and, mainly, for having participated actively in the implementation of modularization in the company, in both Product Engineering and the Production Process. After careful analysis, the elaborated conceptual model was approved by the company's specialists.

7.1.6   Definition of the Time Period of the Analysis

As the research was longitudinal, one of the steps involved determination of the period of analysis. The first aspect that involved such a decision was to determine in which period the project/product analyzed was redesigned and transformed from unmodularized to modularized design/product. When consulting the company's specialists, it was determined that this occurred between September and October 2013.

Thus, the start of the period of analysis was defined as January 2011, and it ended in June 2014. The choice of such a period took into account the fact that data collection was carried out in the months of July, August and September of 2014. The modularized design/product and the non-modularized design/product were analyzed in

CHART 7.1
Company Professionals Consulted

Function	Support for the Project	Service time in the company (years)
Product Engineer	Definition of the model	8
Product Engineer	Definition of the model	13
Team Center Co-ordinator	Definition of the model and the collection of the Product Engineering and Production Process data	10
Product Engineering Manager	Definition of the model	10
Engineering Director	Definition of the model, data collection, validation of the model and interpretation of the results	7
Industrial Director	Definition of the model, data collection, validation of the model and interpretation of the results	7

the same period in the two analysis units defined for study (the Product Engineering and the Production Process).

7.1.7 DEFINITION OF THE DMUs

After determining the analysis period, the DMUs were defined. Support for this process was sought in the literature (e.g. Jain et al., 2011). These authors performed a simulated analysis, and the definition of DMUs indicated important orientations for this analysis. Seeking support in the literature is a procedure we recommend to readers who apply DEA. Jain et al. (2011), for example, determined as DMUs the weekly production batch under analysis in the research. Thus, in Product Engineering, the monthly lot of modularized projects and the monthly lot of non-modularized projects were defined as DMUs. The monthly lot consisted of the total number of projects developed in the period of one month by Product Engineering. In the Production Process, the monthly lot of modularized products and the monthly lot of non-modularized products were defined as DMUs. The monthly lot consisted of the total number of buses produced in the period of one month in the Production Process.

Accordingly, the analyses in the Product Engineering and in the Production Process included 42 DMUs, as follows: (i) 12 DMUs referring to the months of 2011; (ii) 12 DMUs for the months of 2012; (iii) 12 DMUs for the months of 2013; and (iv) six DMUs for the first half of 2014. The DMUs were codified in ascending order (DMU1, DMU2, ... DMUn) to facilitate interpretation of the results and to avoid exposing the company data.

7.1.8 DEFINE THE VARIABLES TO BE USED IN THE DEA MODEL

The process of listing and defining variables was carried out based on the literature (research on modularization, research using DEA) and with the support of the focus group. Subsequently, these listed variables were discussed with the company's specialists. This process was developed integrally for Product Engineering and for the Production Process. For a better understanding, we will show these processes separately.

7.1.8.1 Definition of the Variables of the Model (Inputs and Outputs) of the Product Engineering Analysis

The variables listed for analysis in the Product Engineering and the works that served as the basis are summarized in Chart 7.2.

After the previous elaboration of the variables of the DEA model, each variable was discussed with the company's specialists with the purpose of validating its use and verifying the availability of data to be collected. Initially, the following variables with data available were identified and validated by the company specialists: part numbers, number of items purchased, number of items produced, number of people engaged in Product Engineering, number of technical problems reported, number of product complaints made by customers, and the quantity of projects developed.

However, it was recognized that company data regarding variables described in this initial listing were unavailable. The problem of unavailability of variable data defined in the conceptual model is common in the DEA modeling process. Variables without data available were identified and excluded from the model. These variables are: i) hours worked on the project; ii) compliance with deadlines for project delivery; and iii) project cost. After elimination of the variables with no data available, it was noted that the model did not contain any variable relative to time.

The company's specialists indicated that information on the project lead time (in-transit time, in days, for specification and configuration of the bus design in Product Engineering) was available. Thus, the lead time variable was included in the DEA

CHART 7.2
List of Potential Variables of the Product Engineering Analysis

Variable	Source
Hours worked on the project	Trappey and Chiang (2008)
Part numbers	Chakravarty and Balakrishnan (2007); Napper (2014)
Number of items purchased	Salvador et al. (2002); Chakravarty and Balakrishnan (2007); Napper (2014)
Number of items produced	Salvador et al. (2002); Chakravarty and Balakrishnan (2007); Napper (2014)
Number of persons engaged in the process (Engineering)	Zhu (2000); Duzakin and Duzakin (2007)
Number of technical problems reported	Starr (2010); Jacobs et al. (2011); Feng and Zhang (2014)
Number of product complaints made by customers	Starr (2010); Jacobs et al. (2011); Feng and Zhang (2014)
Projects delivered by the deadline or delayed	Swink et al. (2006)
Cost of the project (US$)	Swink et al. (2006); Trappey and Chiang (2008)
Quantity of projects developed	Trappey and Chiang (2008)

model to represent the time variable. It was also realized that detailed data on this variable could be collected by segregating lead time via processes within the Product Engineering. It was also suggested to include the negotiation lead time (time, in days, from the beginning of the negotiation of the order until its conclusion), because, although such a process is executed by the commercial area, it also involves technical aspects linked to Product Engineering.

Although the opinion of the process specialists was considered relevant, support was sought in the literature to validate the use of the lead time variable in Product Engineering. Supporting literature dealt with research on modularization and assessment of the Product Engineering with the use of DEA. In the research about modularity, it was possible to note that lead time analysis is regarded as important (Joneja & Lee, 2007; Jacobs et al., 2011), including research dealing with bus manufacturers (Lin et al., 2012; Napper, 2014).

Another important aspect is that the process specialists indicated that, in the variable of the quality of projects/products (based on the number of technical problems reported and customers' product complaints), one could also consider the variable of the number of complaints about items, which consists of the number of items of a vehicle that have received customer complaints (e.g. problems with seats, engine, suspension, etc.). As this variable was considered to be significant by the specialists and is described in the literature (Starr, 2010; Jacobs et al., 2011; Feng & Zhang, 2014), it was also included in the model. After these changes, Chart 7.3 shows the list of variables used in Product Engineering. It also shows the function of each variable in the model.

The definitions of the variables as inputs or outputs were made on the basis of the literature (Cook et al., 2014) and in discussions with the process specialists. Cook et al. (2014) pointed out that the resources used in the process being assessed should be defined as inputs and that the results of the use/transformation of these resources should be defined as outputs. After discussion with the process specialists, a consensus was reached that the output to be used in the Product Engineering would be the projects developed. As for the other variables, the consensus was that they should be used as inputs of the DEA model.

7.1.8.2 Definition of the Model Variables (Inputs and Outputs) of the Production Process

As shown, the definition of variables for the Product Engineering and for the Production Process was carried out jointly, considering the same procedures. Initially, a list was made of the variables to be used in the assessment of the Production Process. For this, support was sought from the focus group and the literature. The variables listed for analysis in the Production Process, and the works that served as the basis, are summarized in Chart 7.4.

With the primary listing of possible variables to be considered in the production process, we discussed with the company's specialists in order to validate each variable and assess the availability of data to be collected. Initially, the following variables were identified and validated by the specialists: main materials used in the manufacturing, part numbers, number of items purchased, number of items produced, number of people engaged in production, number of technical problems

CHART 7.3

Final List of Variables of the Product Engineering Analysis

Variable	Source	Function in the model
Commercial lead time (negotiation)	Trappey, Chiang (2008)	Input
Engineering lead time (specification of the order)	Joneja and Lee (2007); Trappey and Chiang (2008); Jacobs et al. (2011); Lin et al. (2012); Napper (2014)	Input
Engineering lead time (product configuration)	Joneja and Lee (2007); Trappey and Chiang (2008); Jacobs et al. (2011); Lin et al. (2012); Napper (2014)	Input
Part numbers	Salvador et al. (2002); Chakravarty and Balakrishnan (2007); Napper (2014)	Input
Number of items purchased	Salvador et al. (2002); Chakravarty and Balakrishnan (2007); Napper (2014)	Input
Number of items produced	Salvador et al. (2002); Chakravarty and Balakrishnan (2007); Napper (2014)	Input
Number of persons engaged in the process (Engineering)	Zhu (2000); Duzakin and Duzakin (2007)	Input
Number of technical problems reported	Starr (2010); Jacobs et al. (2011); Feng and Zhang (2014)	Input
Number of product complaints made by customers	Starr (2010); Jacobs et al. (2011); Feng and Zhang (2014)	Input
Number of item complaints made by customers	Starr (2010); Jacobs et al. (2011); Feng and Zhang (2014)	Input
Number of projects developed	Trappey and Chiang (2008)	Output

reported, number of product complaints made by customers and the number of products manufactured.

With respect to the main materials used in the manufacturing, we tried to define which could be included in the present research. According to guidance from the specialists, those materials of the ones used in the process which were considered on curve A were defined. The most relevant significant important materials (curve A items) in the manufacture of buses were: steel, aluminum, fiber, textiles, fabrics and glass. According to information from the process specialists, these items represented 73% of the total cost of the materials for the product (bus). Thus, these materials (steel, aluminum, fiber, textiles, fabrics and glass) were considered to be variables to be used in the assessment of the Production Process.

It was identified that the following data were not available: working hours, the total unit cost of the products and compliance with deadlines for delivery of the model. The specialists suggested inclusion of the variable lead time, segregated into manufacturing and assembly lead time.

In order to validate the use of the lead time variable in the Production Process, support was sought in the literature from research that considered modularization

---

CHART 7.4

List of Potential Variables of the Production Process

Variable	Source
Main materials used in the fabrication	Jain et al. (2011); Park et al. (2014); Cook et al. (2014)
Hours worked in the production	Jain et al. (2011); Park et al.(2014)
Part numbers	Chakravarty and Balakrishnan (2007); Napper (2014)
Number of items purchased	Salvador et al. (2002); Chakravarty and Balakrishnan (2007); Napper (2014)
Number of items produced	Salvador et al. (2002); Chakravarty and Balakrishnan (2007); Napper (2014)
Number of persons engaged in the process (Production)	Zhu (2000); Duzakin and Duzakin (2007)
Number of technical problems reported	Starr (2010); Jacobs et al. (2011); Feng and Zhang (2014)
Number of product complaints made by customers	Starr (2010); Jacobs et al. (2011); Feng and Zhang (2014)
Orders delivered by the deadline or delayed	Swink et al. (2006); Park et al. (2014)
Total unit cost of the product (US$)	Swink et al. (2006);
Quantity of products manufactured	Jain et al. (2011); Cook et al. (2014)

---

and assessment of the Production Process with the use of DEA. In research on modularization, it was realized that the importance of lead time analysis is considered in studies by Joneja and Lee (2007), Starr (2010), Jacobs et al. (2011), Lin et al. (2012), Napper (2014) and Park et al. (2014).

The process specialists made the same recommendation for the Product Process, regarding the product quality variables, as for Product Engineering. In addition, there was a suggestion to include the variable, number of item complaints, which consists of the number of items of a vehicle that received customer complaints (for example, problems with seats, engine, suspension, etc.). As this variable was considered to be significant by specialists, it was also included in the Production Process model (as it had already been considered in Product Engineering). Thus, it is possible to present the final list of variables defined to be used in the analysis of the Production Process, according to Chart 7.5.

The products manufactured were defined as Outputs of the Production Process. Regarding the other variables, the consensus was that they should be used as inputs to the DEA model.

7.1.9 DEFINITION OF THE DEA (CRS/VRS) MODEL TO BE USED

After choosing the variables to be used in the analysis, the next step was to define the DEA model (CRS or VRS). The model used was CRS, since an internal analysis (internal benchmark) had been carried out in the study of the company under focus. Thus, the scale relationship among the selected variables was similar among the DMUs defined. The use of the CRS model was defined to assess the two units of analysis (the Product Engineering and the Production Process).

CHART 7.5
Final List of Variables in the Production Process

Variable	Theoretical basis	Function in the model
Steel	Jain et al. (2011); Park et al. (2014); Cook et al. (2014)	Input
Aluminum	Jain et al. (2011); Park et al. (2014); Cook et al. (2014)	Input
Fiber	Jain et al. (2011); Park et al. (2014); Cook et al. (2014)	Input
Bus Pass Reader	Jain et al. (2011); Park et al. (2014); Cook et al. (2014)	Input
Fabric	Jain et al. (2011); Park et al. (2014); Cook et al. (2014)	Input
Glass	Jain et al. (2011); Park et al. (2014); Cook et al. (2014)	Input
Production lead time	Joneja and Lee (2007); Starr (2010); Jacobs et al. (2011); Lin et al. (2012); Napper (2014)	Input
Assembly lead time	Joneja and Lee (2007); Starr (2010); Jacobs et al. (2011); Lin et al. (2012); Napper (2014)	Input
Part numbers	Chakravarty and Balakrishnan (2007); Napper (2014)	Input
Number of items purchased	Salvador et al. (2002); Chakravarty and Balakrishnan (2007); Napper (2014)	Input
Number of items produced	Salvador et al. (2002); Chakravarty and Balakrishnan (2007); Napper (2014)	Input
Number of persons engaged in the process (Production)	Zhu (2000); Duzakin and Duzakin (2007)	Input
Number of technical problems reported	Starr (2010); Jacobs et al. (2011); Feng and Zhang (2014)	Input
Number of product complaints made by customers	Starr (2010); Jacobs et al. (2011); Feng and Zhang (2014)	Input
Number of item complaints made by customers	Starr (2010); Jacobs et al. (2011); Feng and Zhang (2014)	Input
Quantity of products manufactured	Jain et al. (2011); Cook et al. (2014)	Output

7.1.10　Definition of the Orientation of the Model (Input or Output)

The input orientation was used, which is recommended when the resources used in the process are more controllable than the outputs (Hamdan & Rogers, 2008). Thus, it is understood that the resources used, both in the Product Engineering and in the Production Process, are more controllable than the number of projects developed, and the number of buses produced (considered the outputs of the models). This orientation was validated by the company's specialists, in view of the fact that the volume of projects developed, and the number of buses produced depends on market demand. The defined input orientation was proposed to reduce the resource use of the inefficient units (Hamdan & Rogers, 2008).

7.1.11 Elaboration and Validation of the Final DEA Model

Therefore, the DEA model project definitions can be presented, and it is possible to present the schematics (Figure 7.3) of the final DEA model of the research. This model was validated by the process specialists.

7.1.12 Collecting Data

Initially, two meetings were held with the focus group that supported the research. At these meetings, guidelines for data collection were discussed. Subsequently, the company's specialists, who had supported the definition of the DEA model, also helped to collect the research data. The definition by the professionals occurred after a 2-day visit to the company. During this visit, a discussion was held with the Engineering Director and the Industrial Director. Engineering and manufacturing processes were also observed, with tours guided by the Product Engineering Coordinator.

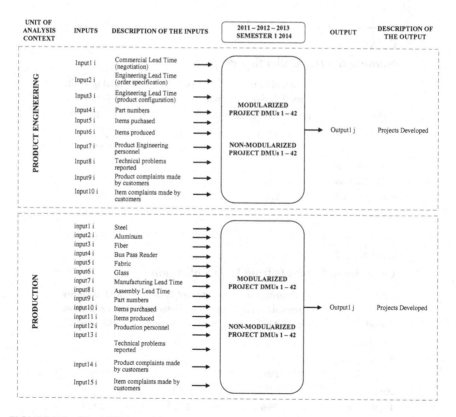

FIGURE 7.3 Final DEA model.

The drafting of the data collection plan was also carried out jointly with the organization's specialists. The data collected on the number of projects are summarized in Chart 7.6.

The data collected in relation to the number of products manufactured are summarized in Chart 7.7.

Data collection is usually the process that requires the most effort on the part of researchers/modelers. In the case illustrated, the data relating to each variable were collected, totaling one variable of the Product Engineering and six variables of the Production Process. This took into account the 946 projects developed and an output of 2,951 products.

7.1.13 HANDLING THE DATA

After collection of all the information necessary for analysis, the data-handling process was started. Initially, the information was organized on spreadsheets. The data were collected in two ways: i) as individual indicators, considering the projects of the Product Engineering and the products manufactured in the Production Process; and ii) as monthly indicators, considering the information generated monthly in the Product Engineering and the Production Process.

CHART 7.6
Number of Data Collection Projects

Period	Quantity of modularized product projects	Quantity of non- modularized product projects	Total quantity of projects in the period
2014 (Jan./Jun.)	49	129	178
2013 (Jan./Dec.)	105	165	270
2012 (Jan./Dec.)	135	110	245
2011 (Jan./Dec.)	140	113	253
General Total	429	517	946

CHART 7.7
Quantity of Products from Data Collection

Period	Quantity of modularized product projects	Quantity of non- modularized product projects	Total quantity of projects in the period
2014 (Jan./Jun.)	211	148	359
2013 (Jan./Dec.)	500	296	796
2012 (Jan./Dec.)	450	300	750
2011 (Jan./Dec.)	645	401	1.046
Total General	1,806	1,145	2,951

The information collected from the monthly indicators did not need to be adjusted, since they were aligned with the definition of the DMU. However, the information collected individually needed to be addressed. To do so, a sum of these data was made according to the date of the projects developed and the number of products manufactured monthly. With regard to Product Engineering, the projects were initially aggregated according to the month of release. For example, if two projects were released in January 2011, they were considered with the batch of projects developed during that month. With the input variables, the same procedure was applied; if the negotiation lead time of project 1 (January 2011, previously considered) was 5 days and that of project 2 was 10 days, the sum of these variables (15 days) was considered as the information. As for the production process, the procedure performed was the same, considering the sum of the individual products of the information collected in the products manufactured monthly.

7.1.14 Carry out the Calculation of Efficiency in DEA

After the modeling process was completed, the previously organized and previously processed data were entered into the application used for the calculation. For the analysis, the composite efficiency was calculated.

7.1.15 Analysis of the Breakdown of the DEA Model Results

In the analysis of the efficiency, it was perceived that the model related to the analysis of the Production Process presented problems of discrimination. Thus, as suggested in the MMDEA, a method for selection of variables (Stepwise) was applied. For illustrative purposes, Stepwise was also applied in the Product Engineering model, even though this model presented no problem of discrimination in the initial analysis.

7.1.16 Use of a Variable Selection Method (Stepwise)

The Stepwise method used in the Product Engineering model was applied in the modularized and non-modularized designs, according to the guidelines of Wagner and Shimshak (2007). Chart 7.8 presents the results of the Stepwise application in the Product Engineering modularized design model:

Initially, the DEA calculation was performed, considering the original model with all input and output variables defined. Subsequently, each variable was removed, and the calculation of the model was performed again. When analyzing the difference between the results of the mean efficiency of the original model, subtracted from the mean efficiency obtained with the exclusion of each variable, it was noticed that no result was equal to 0, meaning that each of the variables had an impact on the efficiency scores of the model proposed. Thus, according to the criteria defined in this study, only variables that did not impact efficiency (mean variation equal to 0) would be excluded from the model, so it was decided that, in the Product Engineering modularized design model, no variables would be excluded. Thus, the model remained as originally established.

CHART 7.8

Stepwise Modularized Design

Representation (E*)	Description of the variable	Mean efficiency (EX*)	Mean efficiency variation (EX*0–EX*n)
E*0 Modularized project	Original model (all the variables)	0.6138	–
E*1 Remove input1 i	Commercial lead time (negotiation)	0.6691	–0.0553
E*2 Remove input2 i	Engineering lead time (order specification)	0.6563	–0.0425
E*3 Remove input3 i	Engineering lead time (product configuration)	0.6888	–0.0750
E*4 Remove input4 i	Part numbers	0.6694	–0.0556
E*5 Remove input5 i	Items purchased	0.6961	–0.0824
E*6 Remove input6 i	Items produced	0.6691	–0.0553
E*7 Remove input7 i	Product Engineering personnel	0.6694	–0.0556
E*8 Remove input8 i	Technical problems reported	0.6723	–0.0586
E*9 Remove input9 i	Product complaints made by customers	0.6681	–0.0543
E*10 Remove input10 i	Item complaints made by customers	0.6733	–0.0595

Subsequently, the same calculation explained in the DEA model was performed for the Product Engineering non-modularized design model. The results are summarized in Chart 7.9.

The analysis of Chart 7.9 also shows that all defined variables impacted on efficiency, so no variables were excluded from the model analysis related to the Product Engineering non-modularized design. Thus, it was defined that the DEA model developed for execution of the Product Engineering analysis would not undergo any alteration.

For the Stepwise analysis of the Production Process, the same procedure was followed for the Product Engineering analysis, with the method being applied to both the modularized and the non-modularized product models. Chart 7.10 presents the results of applying Stepwise to the modularized product model of the Production Process.

When analyzing the difference between the results of the mean efficiency of the original model and the mean efficiency obtained with the exclusion of each variable, it can be seen that three variables presented variation equal to 0. These variables could be removed from the model, since they did not impact the efficiency scores of the DEA model developed. These variables were: (i) steel; (ii) part numbers; and (iii) item complaints made by customers. In compliance with the criteria defined in the research, these variables were excluded from the model.

CHART 7.9
Stepwise Non-Modularized Design

Representation (E*)	Description of the variable	Mean efficiency (EX*)	Mean efficiency variation (EX*0–EX*n)
E*0 Non-modularized project	Original model (all variables)	0.5115	–
E*1 Remove input1 i	Commercial lead time (negotiation)	0.4728	0.0387
E*2 Remove input2 i	Engineering lead time (order specification)	0.4808	0.0307
E*3 Remove input3 i	Engineering lead time (product configuration)	0.4818	0.0297
E*4 Remove input4 i	Part numbers	0.4807	0.0308
E*5 Remove input5 i	Items purchased	0.4724	0.0391
E*6 Remove input6 i	Items produced	0.4808	0.0307
E*7 Remove input7 i	Product Engineering personnel	0.4808	0.0307
E*8 Remove input8 i	Technical problems reported	0.5650	0.0535
E*9 Remove input9 i	Product complaints made by customers	0.4838	0.0277
E*10 Remove input10 i	Item complaints made by customers	0.4758	0.0357

Subsequently, the same calculation explained in the DEA model for the Production Process non-modularized model was performed, and the results are summarized in Chart 7.11.

After performing the procedures related to Stepwise of the non-modularized product in the Production Process, it was noticed that the method suggested the exclusion of four variables: (i) steel; (ii) part numbers; (iii) production personnel; and (iv) item complaints made by customers. As observed, the exclusion of three of these variables (steel, part numbers and item complaints made by customers) was also suggested in the previous analysis regarding the modularized product. The variable related to production personnel was not excluded from the modularized product model, although the Stepwise method suggested that it be eliminated from the non-modularized product model. According to Dyson et al. (2001), one of the main assumptions for the application of Data Envelopment Analysis (DEA) is that the defined set of variables should be common to all the units of analysis. Thus, it was decided not to exclude this variable (production personnel) from the analysis performed.

After deleting the variables suggested by the Stepwise method, the reorganized, processed data were re-inserted into the application used for analysis. The results obtained showed that the problem of discrimination was solved, allowing the data analysis to proceed. As the use of Stepwise showed the need to exclude input

CHART 7.10

Stepwise Modularized Product

Representation (E*)	Description of the variable	Mean efficiency (EX*)	Mean efficiency variation (EX*0–EX*n)
E*0 Modularized product	Original model (all variables)	0.4988	–
E*1 Remove input1 i	Steel	0.4988	0.0000
E*2 Remove input2 i	Aluminum	0.5047	–0.0059
E*3 Remove input3 i	Fiber	0.4975	0.0013
E*4 Remove input4 i	Bus Pass Reader	0.4980	0.0008
E*5 Remove input5 i	Fabric	0.4991	–0.0003
E*6 Remove input6 i	Glass	0.5048	–0.0060
E*7 Remove input7 i	Lead time Fabrication	0.4993	–0.0005
E*8 Remove input8 i	Lead time Assembly	0.4985	0.0003
E*9 Remove input9 i	Part numbers	0.4988	0.0000
E*10 Remove input10 i	Items purchased	0.4988	–0.0001
E*11 Remove input11 i	Items produced	0.4991	–0.0003
E*12 Remove input12 i	Production personnel	0.4987	0.0001
E*13 Remove input13 i	Technical problems reported	0.5015	–0.0027
E*14 Remove input14 i	Product complaints made by customers	0.4986	0.0002
E*15 Remove input15 i	Item complaints made by customers	0.4988	0.0000

variables in the DEA model originally developed, adjustments were made to the final DEA model. The adjusted final DEA model is shown in Figure 7.4.

It should be noted that the final models of both the Product Engineering and the Production Process analyses respected the conditions indicated in the literature, that the number of DMUs should be at least three times higher than the number of inputs and outputs added together (Banker et al., 1989; Cook et al., 2014).

7.1.17 ANALYSIS OF THE RESULTS (EFFICIENCY, TARGETS AND SLACK AND BENCHMARKS)

For analysis of the results (which will be detailed in Chapter 10), graphs and charts were prepared to better visualize the evolution of efficiency.

7.1.18 PRESENTATION AND DISCUSSION OF THE FINAL RESULTS

With the results obtained and the analyses carried out, a meeting was held to present and discuss the results with the process specialists. This procedure was important for

CHART 7.11

Stepwise Non-Modularized Product

Representation (E*)	Description of the variable	Mean efficiency (EX*)	Mean efficiency variation. (EX*0–EX*n)
E*0 Modularized product	Original model (all variables)	0.5022	–
*E*1 Remove input1 i*	*Steel*	*0.5022*	*0.0000*
E*2 Remove input2 i	Aluminum	0.5027	–0.0005
E*3 Remove input3 i	Fiber	0.5017	0.0005
E*4 Remove input4 i	Bus Pass Reader	0.5024	–0.0003
E*5 Remove input5 i	Fabric	0.5020	0.0002
E*6 Remove input6 i	Glass	0.5023	–0.0001
E*7 Remove input7 i	Lead time Fabrication	0.5026	–0.0004
E*8 Remove input8 i	Lead time Assembly	0.5037	–0.0015
*E*9 Remove input9 i*	*Part numbers*	*0.5022*	*0.0000*
E*10 Remove input10 i	Items purchased	0.5018	0.0003
E*11 Remove input11 i	Items produced	0.5021	0.0001
*E*12 Remove input12 i*	*Production personnel*	*0.5022*	*0.0000*
E*13 Remove input13 i	Technical problems reported	0.5020	0.0001
E*14 Remove input14 j	Product complaints made by customers	0.5021	0.0001
*E*15 Remove input15 j*	*Item complaints made by customers*	*0.5022*	*0.0000*

a better understanding of the results obtained. We recommend that such a meeting be recorded or filmed so that it helps the researcher/modeler to revisit the specialists' settings for better data analysis.

7.1.19 DEFINITION OF THE IMPROVEMENT GOALS, BASED ON THE RESULTS OBTAINED WITH THE ANALYSIS

Together with the company's specialists, an action plan was drawn up to seek the improvements suggested by the analyses which had been carried out.

7.1.20 WRITING THE FINAL REPORT

The final report was drafted, resulting in a master's dissertation written by one of the authors of this book.

7.2 APPLICATION OF MMDEA IN A SERVICE PROVISION SYSTEM

In the sequence, we will show the application of the proposed method in an analysis performed in a service provision system. The application presented refers to the

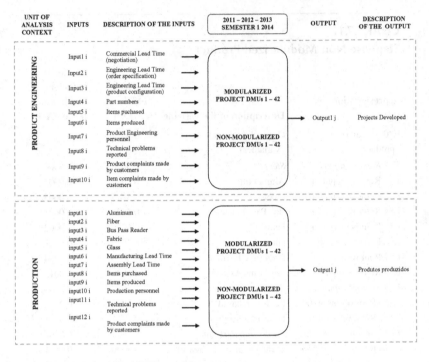

FIGURE 7.4   Adjusted final DEA model.

practical problem of efficiency analysis presented in Chapter 3, Section 3.2. The focus will be on the application of the modeling method, the results of which will be shown in Chapter 10, Section 10.2.

7.2.1   DEFINITION OF THE PURPOSE OF THE ANALYSIS AND THE TYPE OF EFFICIENCY TO BE ASSESSED

The area of operations may contribute to the competitiveness of service companies by increasing productivity and efficiency of their processes. Seeking to contribute in this way, the research theme is focused on service operations. The focus of the analysis has the following general objective:

- Analyze which are the variables present in service contracts that affect efficiency in service operations.

The specific objectives of the work are:

- Establish an efficiency metric of service contracts that is considered from the perspective of the customer, service provider and with both integrated.
- Identify whether there are significant differences among the efficiency scores of service contracts from the perspective of the client, service provider and with both integrated.

- Carry out an analysis of the prevalent variables in the efficiency of service provision.

It was defined that the types of efficiency to be analyzed were the following three with respect to technical efficiency: standard, inverted frontier and composite. In some analyses, only the composite technical efficiency was used.

7.2.2 CONDUCT OF A SYSTEMATIC REVIEW OF THE LITERATURE

The literature review was carried out to identify research involving the application of DEA to the measurement of efficiency in service operations. In order to carry out this research, we used the keywords "service operations", combined with "data envelopment analysis", "productivity", "efficiency" and "performance". The research was carried out with these combinations of words in both English and Portuguese.

7.2.3 DEFINITION OF THE UNITS OF ANALYSIS

It was agreed that the research scope would include management of the maintenance service of automotive fleets. This service was chosen by adherence to the research objective and relevance for the company studied. The service is considered to be adherent to the research, since it presents variables that can influence the efficiency of the operation, such as the characteristics of leased fleets and client service contracts. In addition, it was felt that the constant interaction of clients with processes can affect the productivity and efficiency of the operation. After defining the service, it was established that service contracts constituted units of analysis.

7.2.4 DEVELOPMENT OF A DEA CONCEPTUAL MODEL

After the systematic review of the literature, and with the support of a focus group, it was possible to identify research that helped with the development of the DEA conceptual model. Subsequently, the research design was developed (Figure 7.5).

The research design shown in Figure 7.5 is composed of two blocks: context analysis and object of study. The context analysis refers to the operation of service contracts of the company studied. The operation of service contracts has three stages: parameterization of the service contract, execution of the contract and support for the contract. The parameterization is the stage where the parameters for the execution of the contract are defined in conjunction with the client. Among the main parameters defined are the periodicity of the service provision, scope of service use, estimate of the number of transactions and estimate of the services that will be rendered.

The stage of execution of the contract is where the use of the service by the client occurs according to the parameterization performed. Therefore, the establishment of the rules of parameterization of the contract impacts the execution of the operation. Support for the contract serves as a client support to use the service continuously. The performance of this step is related to the previous steps, in that the better the parameterization and execution of the service, the less the need to use the support.

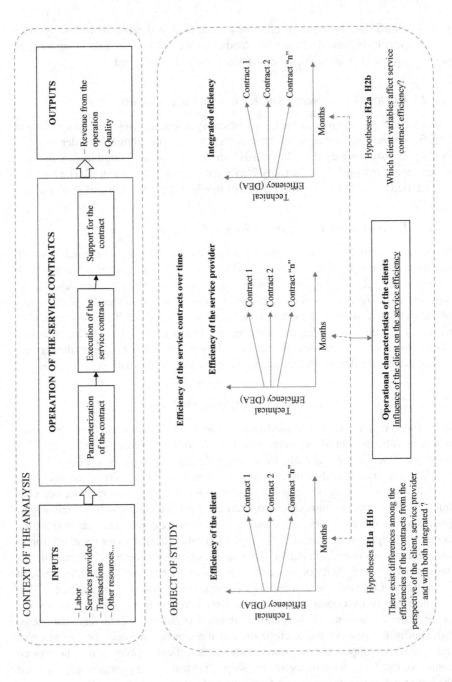

FIGURE 7.5 Design of the research analysis in the service company.

The operation of each service contract is composed of inputs that are processed and transformed into outputs. The inputs can be considered as the inputs needed for the contract to be fulfilled. In this case, they are related to: (a) the workforce required to execute the contract; (b) the quantity of services provided (via telephone calls and email, requests for services, etc.); (c) transactions, that is, the number of times the service is used; and (d) resources related to infrastructure, information, etc. The output can be considered to be the result of the contract fulfillment process, which is the revenue obtained by the company to meet the contract, and quality, which is related to the service level agreed with the client in the contract.

The analysis of the technical efficiency of contracts over time was carried out from three different perspectives, namely the perspective of the client, of the service provider and with both integrated. Efficiency from the client perspective aims to demonstrate the efficiency of service delivery contracts based on client-related variables. Efficiency from the perspective of the service provider aims to demonstrate the efficiency of service provision contracts based on variables related to the company under study. Integrated efficiency comprises the joint use of client and service provider variables.

7.2.5 ASSESSMENT OF THE DEA CONCEPTUAL MODEL

Specialists from the company were invited to assess the DEA conceptual model that was developed. Chart 7.12 presents the specialists involved in the process, as well as their responsibility, service time and training.

The process specialists were chosen for their knowledge of the operation analyzed, company service time, recurring contact with clients, and knowledge about service contracts. The model was considered to be satisfactory and was approved by the process specialists.

CHART 7.12
Company Professionals Consulted

Function	Responsibility	Service time in company
Manager of operations	Support for definition of the DEA model. Authorization for data collection.	10 years
Co-ordinator of services	Support for definition of the DEA model. Support for data collection.	5 years
Analyst of processes	Support for definition of the DEA model. Support for data collection.	2 years
Co-ordinator of client service	Support for definition of the DEA model. Support for data collection.	6 years
Analyst of financial control	Support for definition of the DEA model. Support for data collection.	2 years
Senior consultant	Support for analysis of results.	5 years

7.2.6 Definition of the Time Period of the Analysis

A longitudinal analysis was defined. From this definition, it became necessary to define the time period of the analysis. Thus, the mean duration of the client contracts was assessed. From this assessment, the following assumptions were identified and should be considered in the definition of the period of analysis: a) the minimum contract time of a client is 12 months, which can be renewed automatically for another 12 months; b) as each client has specific start and finish dates, it was not possible to establish a standard in the total time span of contract duration; and c) the company's ERP system, responsible for storing the information, was implemented at the end of 2014. Based on these premises, the time interval from January to December 2015 (one year) was defined as the period proposed for the analysis.

7.2.7 Definition of the DMUs

After establishing the analysis period, the DMUs of the model were defined. It was initially considered that each DMU would be composed of the service contract and the month of analysis. Based on this definition, the process specialists warned about the existence of specific contracts for certain clients. Specific contracts are characterized by specific business or operational conditions for the service. To overcome these anomalies, it was necessary to carry out assessment of such contracts with their specific characteristics. The specific contracts identified were removed from the research universe and were not considered in the efficiency assessments, since they would make it impossible to execute DEA due to the lack of homogeneity of the DMUs.

Following the exclusion of the specific contracts, nine contracts were selected that had their active service in the period from January to December 2015. The combination of the contracts selected with the analysis time interval resulted in 108 DMUs. To assist with the traceability of the DMUs throughout the execution of the model, a codification was defined, as shown in Figure 7.6. In this figure, it is possible to identify, for example, that the DMU called C1P1 is the combination of the contract number 1 (C1) with its analysis period 1 (P1) (January 2015).

7.2.8 Definition of the Variables to Be Used in the DEA Model

The process of defining the variables was started by means of a listing drawn from the systematic literature review. The work was analyzed with the objective of identifying the inputs and outputs currently used in research with data envelopment analysis in service operations. Chart 7.13 presents the variables identified in the literature.

The variables contained in Chart 7.13 were presented to the process specialists to identify the viability of their use. The process specialists validated the following variables for the DEA model: (a) capacity; (b) transactions; (c) sales volume; (d) revenue; and (e) service quality. The transaction variable was categorized into five variables: (a) quantity of service orders generated per client; (b) number of clients contacted by telephone; (c) number of email exchanges with clients; (d) number of clients directed to workshops; and (e) number of vehicle tests. The revenue variable was categorized into service revenue and US$ per km.

CONTRACT

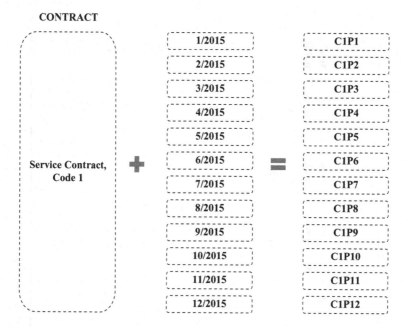

FIGURE 7.6 Illustration of DMUs.

The variables, number of employees, operating costs and hours worked were not considered in the model, since, as the objective is to assess the efficiency of each client contract, it was not possible to apportion the number of employees, operating costs and hours worked among individual client contracts. Apportionment was impractical, since a single employee could be involved in more than one client contract, and the number of hours allocated to rendering the service to each client is not registered by the company studied in its ERP.

The classification of the variables as inputs and outputs was based on the assessment by the process specialists and the definition by Cook et al. (2014). Chart 7.14 presents the variables selected for the DEA model, with classification of inputs and outputs according to the perspective in which the efficiency was being assessed.

7.2.9 DEFINITION OF THE DEA (CRS/VRS) MODEL TO BE USED

The use of VRS was defined because the scale relationship among the selected variables is different among the DMUs.

7.2.10 DEFINITION OF THE MODEL (INPUT OR OUTPUT) ORIENTATION

We chose to use input orientation, since it is understood that the inputs used to render the services are more controllable than the outputs. Some of the outputs defined, such as revenue, are set out in the contract, and the area of operations does not have the autonomy to interfere with this process.

CHART 7.13

Variables Proposed in the Analysis and the Base References

References	No. of employees	Hours of work	Capac.	Operational costs	Transactions	Sales volume	Revenue	Service quality
Akhtar (2010)	X		X				X	
Cook and Zhu (2005)					X	X		
Donthu and Yoo (1998)						X		
Kantor and Maital (1999)					X			
Khaira (2008)				X			X	
Lin and Huang (2009)							X	
Lorenzo and Sanchez (2007)	X		X	X				
Liu and Li (2014)	X						X	
O'Neill and Dexter (2004)					X			
Resende and Tupper (2009)			X	X				X
Shang et al. (2008)	X		X	X				
Sherman and Zhu (2006)				X				
Shimshak and Lenard (2007)								X
Soteriou and Zenios (1999)		X						X
Staat (2006)					X			
Yunshi and Chich-Jen (2011)			X				X	
Zervopoulos and Palaskas (2011)	X	X	X					

CHART 7.14

Final Variables and Function Definition in the Model

Reference	Variable	Unit	Client perspective	Service provider perspective	Integrated perspective
Capacity	Quantity of vehicles	Quantity	Input	Input	Input
Transactions	Service orders	Quantity	Input	Input	Input
Transactions	Clients attended by tel.	Quantity	Input	Input	Input
Transactions	Clients attended by email	Quantity	Input	Input	Input
Transactions	Directions	Quantity	Input	Input	Input
Transactions	Tests	Quantity	Input	Input	Input
Sales volume	Maintenance volume	US$	Input	Output	Output
Revenue	Service revenue	US$	Input	Output	Output
Revenue	US$ per km	US$	Output	Not applicable	Output
Service quality	Contract period	Days	Not applicable	Output	Output

7.2.11 ELABORATION AND VALIDATION OF THE FINAL DEA MODEL

After establishing the DEA model definitions, it is possible to present the schematics (Figure 7.7) of the model defined for this research. The final DEA model was validated by the process specialists.

7.2.12 COLLECTING DATA

The first stage of this phase was to carry out data collection planning. This planning took place through a meeting with the specialists in the process, with the objective of defining: a) the data collection sources; b) the data collection period; and c) the forms of encryption or masking of the data to preserve the confidentiality of the information about the company. The data sources suggested for the DEA model are presented in Chart 7.15.

After the information was collected to compose the DEA model variables, the operational characteristics of the clients' fleets were collected together with the process specialists. The operational characteristics of client fleets are associated with client influence in the delivery service process. The first mapped feature was the type of item used in the fleet. The type of network used was the second feature mapped with the process specialists. The region of action of the fleet was the third characteristic informed by the specialists. The mean age of the fleet was the fourth characteristic identified by the specialists. The fifth mapped feature was the vehicle family (car, motorcycle or truck). The last mapped feature was preventive maintenance of the fleet.

7.2.13 HANDLING THE DATA

After collecting all the information necessary for analysis, the data were organized in spreadsheets. In addition, the data was masked so that information considered to be confidential by the organization under study was not exposed.

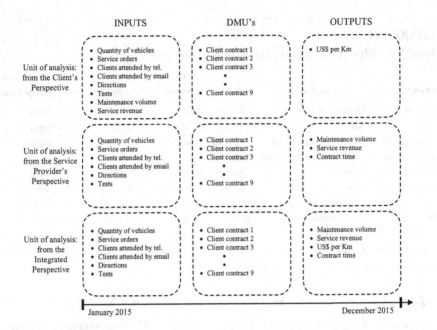

FIGURE 7.7   Final DEA model.

CHART 7.15

Data Collection Sources

Source	Information collected
SAP System	Volume of client expenditure on fleet maintenance
	Service revenue
CRM	Quantity of vehicles in clients' fleets
	Service contract records
	Contract period
Internal System	Service orders
	Directions to workshops
	Vehicle tests
	Vehicle mileage
Call Center Sistema	Clients contacted via telephone
	Clients contacted via email

7.2.14   CARRYING OUT THE EFFICIENCY CALCULATION IN DEA

After the modeling process was complete, the previously organized and processed data were entered into the application used for the calculation.

7.2.15   ANALYSIS OF THE BREAKDOWN OF THE DEA MODEL RESULTS

It was observed that the models concerning the perspectives of the Service Provider and the Integrated Service presented problems of discrimination. Thus, as suggested in the MMDEA, a method for selection of variables (Stepwise) was applied. For illustrative purposes, Stepwise was also applied in the model regarding the client perspective, which did not present a discriminatory problem.

7.2.16   USING A VARIABLE SELECTION METHOD (STEPWISE)

As recommended by Wagner and Shimshak (2007), the Stepwise method was initially performed considering all the input and output variables for the efficiency models from the client, service provider and integrated perspectives. When analyzing the difference in the mean efficiency results of the original model and the mean efficiency obtained with the exclusion of each variable in the model from the perspective of the client, it was identified that all variables generated significant impacts on the efficiency scores. Thus, it was defined that, in the model from the perspective of the client, all the original variables would be considered.

For the models from the service provider and integrated perspectives, variables that did not generate significant impacts on the mean of efficiency and that impaired the discrimination of the efficiency scores were identified, namely service orders, testing and volume of maintenance. The contract period was relevant for the analysis from the perspective of the service provider, but not for the analysis from an integrated perspective. These variables were removed from the models. Chart 7.16 shows the eliminated variables and those that were maintained in each of the analysis models (client, service provider and integrated perspectives).

After excluding the variables suggested by the Stepwise method, the reorganized processed data were re-inserted into the application used for the calculation. The results obtained showed that the discrimination problem was solved, allowing the data analysis to proceed. As the use of Stepwise showed the need to exclude input variables in the DEA models originally developed, adjustments were made to the final DEA model, as shown in Figure 7.8.

CHART 7.16
Source: Final List of the Analysis Model Variables

Variable	Client	Service provider	Integrated
Quantity of vehicles	Maintained	Maintained	Maintained
Service orders	Maintained	Excluded	Excluded
Clients contacted by tel.	Maintained	Maintained	Maintained
Clients contacted by email	Maintained	Maintained	Maintained
Directions	Maintained	Maintained	Maintained
Tests	Maintained	Excluded	Excluded
Maintenance volume	Maintained	Excluded	Excluded
Service revenue	Maintained	Maintained	Maintained
US$ per km	Maintained	Not applicable	Maintained
Contract period	Maintained	Maintained	Excluded

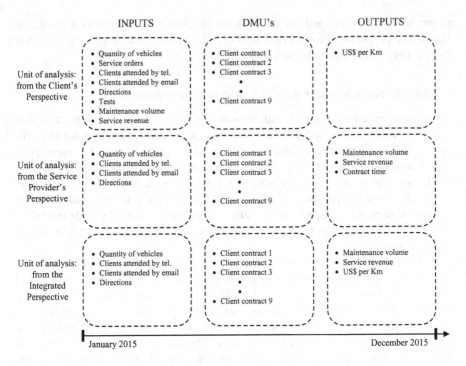

FIGURE 7.8   Adjusted final DEA model.

7.2.17 ANALYSIS OF THE RESULTS (EFFICIENCY, TARGETS AND SLACK AND BENCHMARKS)

For analysis of the results, the following procedures were executed: (i) graphical analysis of the efficiency behavior over time, (ii) elaboration of analysis charts, (iii) statistical tests to identify significant differences among the mean efficiency scores. The analyses can be found in Chapter 10, Section 10.2.

7.2.18 PRESENTATION AND DISCUSSION OF THE FINAL RESULTS

With the results obtained and analyses carried out, a meeting was held to present and discuss the results with the process specialists. This procedure proved to be important and helped to gain a better understanding of the results.

7.2.19 DEFINITION OF THE IMPROVEMENT GOALS, BASED ON THE RESULTS OBTAINED WITH THE ANALYSIS

Together with the company's specialists, an action plan was drawn up to achieve the improvements suggested by the analyses.

7.2.20 WRITING OF THE FINAL REPORT

Lastly, the final report was drafted, leading to a master's dissertation.

REFERENCES

Akhtar, M. H. (2010). X-Efficiency analysis of Pakistani commercial banks. *International Management Review, 6*(1), 12.

Banker, R. D., Charnes, A., Cooper, W. W., Swarts, J., & Thomas, D. (1989). An introduction to data envelopment analysis with some of its models and their uses. *Research in Governmental and Nonprofit Accounting, 5*(1), 125–163.

Chakravarty, A.K., & Balakrishnan, N. (2007). Achieving product variety through optimal choice of module variations. *IIE Transactions, 33*(7), 587–598.

Cook, W. D., & 2 Zhu, J. O. E. (2005). Building performance standards into data envelopment analysis structures. *IIE Transactions, 37*(3), 267–275.

Cook, W., Tone, K., & Zhu, J. (2014). Data envelopment analysis: Prior to choosing a model. *Omega, 44*(1), 1–4.

Donthu, N., & Yoo, B. (1998). Retail productivity assessment using data envelopment analysis. *Journal of Retailing, 74*(1), 89–105.

Duzakin, E., & Duzakin, H. (2007). Measuring the performance of manufacturing firms with super slacks-based model of data envelopment analysis: An application of 500 major industrial enterprises in Turkey. *European Journal of Operational Research, 182*(3), 1412–1432.

Dyson, R. G., Allen, R., Camanho, A. S., Podinovski, V. V., Sarrico, C. S., & Shale, E. A. (2001). Pitfalls and protocols in DEA. *European Journal of Operational Research, 132*(2), 245–259.

Feng, T.,& Zhang, F. (2014). The impact of modular assembly on supply chain efficiency. *Production and Operations Management, 23*(11), 1985–2001.

Hamdan, A., & Rogers, K.J. (2008). Evaluating the efficiency of 3PL logistics operations. *International Journal of Production Economics, 113*(1), 235–244.

Jacobs, M., Droge, C., Vickery, S.K., & Calantone, R. (2011). Product and process modularity's effects on manufacturing agility and firm growth performance. *Journal of Product Innovation Management, 28*(1), 123–137.

Jain, S., Triantis, K. P., & Liu, S. (2011). Manufacturing performance measurement and target setting: A data envelopment analysis approach. *European Journal of Operational Research, 214*(3), 616–626.

Joneja, A., & Lee, N. (2007). A modular, parametric vibratory feeder: A case study for flexible assembly tools for mass customization. *IIE Transactions, 30*(10), 923–931.

Kantor, J., & Maital, S. (1999). Measuring efficiency by product group: Integrating DEA with activity-based accounting in a large mideast bank. *Interfaces, 29*(3), 27–36.

Khaira, R., (2008) *Using segment attractiveness to improve segment selection in the credit card business.* American Marketing Association.

Lin, L., & Huang, C. Y. (2009). Optimal size of the financial services industry in Taiwan: A new DEA-option-based merger simulation approach. *The Service Industries Journal, 29*(4), 523–537.

Lin, Y., Ma, S., & Zhou, L. (2012). Manufacturing strategies for time based competitive advantages. *Industrial Management & Data Systems, 112*(5), 729–747.

Liu, J., & Li, W. (2014). Efficiency measures of the internet company in china using a three-stage DEA model. *Pakistan Journal of Statistics, 30*(5).

Lorenzo, J. M. P., & Sánchez, I. M. G. (2007). Efficiency evaluation in municipal services: An application to the street lighting service in Spain. *Journal of Productivity Analysis, 27*(3), 149–162.

Napper, R. (2014). Modular route bus design–A method of meeting transport operation and vehicle manufacturing requirements. *Transportation Research Part C: Emerging Technologies, 38*(1), 56–72.

O'neill, L., & Dexter, F. (2004). Market capture of inpatient perioperative services using DEA. *Health Care Management Science, 7*(4), 263–273.

Park, J., Lee, D., & Zhu, J. (2014). An integrated approach for ship block manufacturing process performance evaluation: Case from a Korean shipbuilding company. *International Journal of Production Economics, 156*(1) 214–222.

Resende, M., & Tupper, H. C. (2009). Service quality in Brazilian mobile telephony: An efficiency frontier analysis. *Applied Economics, 41*(18), 2299–2307.

Salvador, F., Forza, C., & Rungtusanatham, M. (2002). Modularity, product variety, production volume, and component sourcing: theorizing beyond generic prescriptions. *Journal of Operations Management, 20*(5), 549–575.

Shang, J. K., Hung, W. T., & Wang, F. C. (2008). Service outsourcing and hotel performance: Three-stage DEA analysis. *Applied Economics Letters, 15*(13), 1053–1057.

Sherman, H. D., & Zhu, J. (2006). Benchmarking with quality-adjusted DEA (Q-DEA) to seek lower-cost high-quality service: Evidence from a US bank application. *Annals of Operations Research, 145*(1), 301–319.

Shimshak, D. G., & Lenard, M. L. (2007). A two-model approach to measuring operating and quality efficiency with DEA. *INFOR: Information Systems and Operational Research, 45*(3), 143–151.

Soteriou, A., & Zenios, S. A. (1999). Operations, quality, and profitability in the provision of banking services. *Management Science, 45*(9), 1221–1238.

Staat, M. (2006). Efficiency of hospitals in Germany: A DEA-bootstrap approach. *Applied Economics, 38*(19), 2255–2263.

Starr, M.K. (2010). Modular production: 45 years hold concept. *International Journal of Operation and Production Management, 30*(1), 7–19.

Swink, M., Talluri, S., & Pandejpong, T. (2006). Faster, better, cheaper: A study of NPD project efficiency and performance tradeoffs. *Journal of Operations Management, 24*(5), 542–562.

Trappey, A., & Chiang, T. (2008). A DEA benchmarking methodology for project planning and management of new product development under decentralized profit-center business model. *Advanced Engineering Informatics, 22*(4), 438–444.

Wagner, J. M., & Shimshak, D. G. (2007). *Stepwise* selection of variables in data envelopment analysis: Procedures and managerial perspectives. *European Journal of Operational Research, 180*(1), 57–67.

Yunshi, M., & Chich-Jen, S. (2011). A study on service quality performance in catering industry – the application of DEA. *Pakistan Journal of Statistics, 27*(5), 573–580.

Zervopoulos, P., & Palaskas, T. (2011). Applying quality-driven, efficiency-adjusted DEA (QE-DEA) in the pursuit of high-efficiency–high-quality service units: An input-oriented approach. *IMA Journal of Management Mathematics, 22*(4), 401–417.

Zhu, J. (2000). Multi-factor performance measure model with an application to Fortune 500 companies. *European Journal of Operational Research, 123*(1), 105–124.

8 Application for Calculation of Productivity and Efficiency in DEA

This chapter shows how to operate an application that can be used to perform productivity and efficiency calculations, using the DEA technique. The application is SAGEPE (System for Analysis and Management of Productivity and Efficiency), developed by the authors of this book.

8.1 SAGEPE APPLICATION

The SAGEPE application can be found at http://www.sagepe.com.br/, APP handle. Initially, it is necessary to organize the data in a table constructed as an Excel spreadsheet (for example, Chart 8.1). The DMUs should be described in the rows and the variables in the columns. Thus, it is possible to organize the data of each variable for each DMU. In the example shown in Chart 8.1, an analysis performed among 10 Brazilian ports is shown. In this case, each port is considered to be a DMU. The input variables selected are: (i) the number of cranes in each port; (ii) the number of berths for mooring vessels in each port; (iii) the area of each port in square meters; and (iv) the number of employees in each port. In addition, the total volume of cargo handled (TEU) per port is considered as the output variable. In this example, a cross-sectional analysis was performed with 10 distinct DMUs over the same time period.

After organizing the data, SAGEPE can be opened and the file loaded into the application. It is to be emphasized that the file to be inserted into SAGEPE should be saved in the format csv or xml. In addition, it is possible to download a guiding template from a data organization model on the application page. The application offers options for importing data, model settings, and results. Furthermore, in the "About" handle, there is additional information about the application. To import the data, clicking on Browse allows selection of the file with the processed data, as displayed in Figures 8.1, 8.2, and 8.3.

After uploading the data, this information can be viewed by clicking on "View imported data". In this case, the application will show a list with the same data organized as shown in Chart 8.1.

With the data entered into the application, the configuration part of the model (Figure 8.4) can be started. We emphasize that the application does not have restrictions regarding the number of variables and DMUs to be used in the analyses.

CHART 8.1
Example of Data Processed on a SAGEPE Spreadsheet

DMU	Cranes	Berths	Area (m²)	Employees	Cargo handled (TEUs)
Suape	2.00	3.00	270.00	120.00	110.00
Salvador	3.00	2.00	74.00	250.00	106.00
Vila Velha	4.00	2.00	100.00	273.00	123.00
Multi-Rio	2.00	2.00	180.00	257.00	151.00
Libra T1	3.00	2.00	155.00	220.00	119.00
Sepetiba	4.00	2.00	400.00	238.00	21.00
Santos	6.00	3.00	366.00	500.00	432.00
Libra T37	5.00	5.00	220.00	620.00	470.00
Paranaguá	2.00	2.00	210.00	149.00	250.00
Rio Grande	4.00	2.00	200.00	414.00	444.00

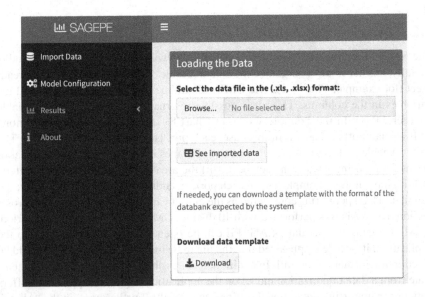

FIGURE 8.1   SAGEPE home screen.

The first step in model configuration is the selection of the variables to be used. Initially, the inputs can be selected and then the outputs. Note that the selection of the inputs (cranes, berths, areas and employees) and the output (TEUs) are exactly as defined in Chart 8.1. The application describes where to make these selections. Finally, the orientation (input v. output) and model type (CRS v. VRS) can be selected. In the example illustrated, the CRS model, with input orientation, is selected. The configuration of the model exemplified is shown in Figure 8.5.

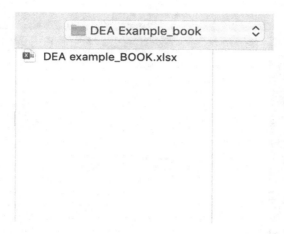

FIGURE 8.2   Example of loading data into SAGEPE.

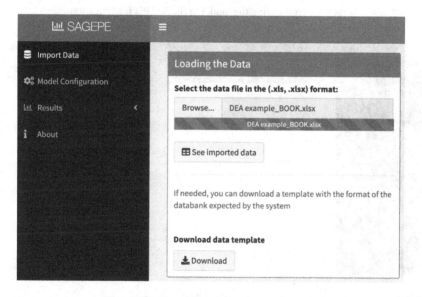

FIGURE 8.3   Data loaded into the SAGEPE application.

After importing the data and settings of the model, analysis of the results can be performed. To analyze the results, the button "Results" is clicked, and the desired analysis selected. In the analysis of the results it is possible to verify the following efficiencies: standard technical, inverted, composite, normalized composite and scale. In addition, it is possible to check slack, benchmarking and, if necessary, to perform Stepwise.

In the example illustrated, we will use standard technical efficiency for analysis. As for the standard technical efficiency results (Figure 8.6), it is possible to observe that the DMUs for the Paranaguá and Rio Grande ports are the most efficient in the

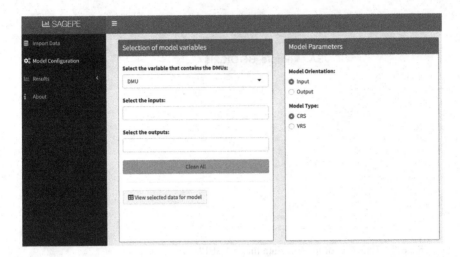

FIGURE 8.4 Screen for configuring the model in the SAGEPE application.

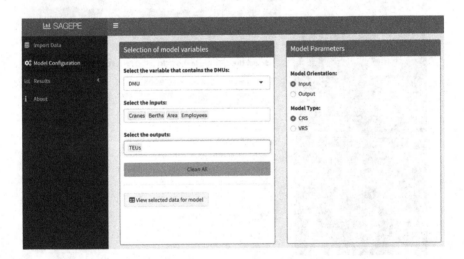

FIGURE 8.5 Sample model configured in the SAGEPE application.

set analyzed (Eff. Standard = 1, or 100%). In contrast, the port of Sepetiba has the worst result (Standard Eff = 0.07, or 7%). It can also be noted that the analysis model developed presents a clear discrimination, not requiring, in this case, execution of the steps for selection of variables, using Stepwise.

Considering the analysis with the standard technical efficiency, it is possible to determine the slack that each DMU has in each variable. This slack represents the opportunities to reduce the use of resources in each variable. For example (Figure 8.7), in order for the port of Suape to become efficient, it needs to eliminate one crane (the values in this interpretation were rounded up or down). In addition, it is possible to eliminate two berths, reduce the area by 177.60 m² and make

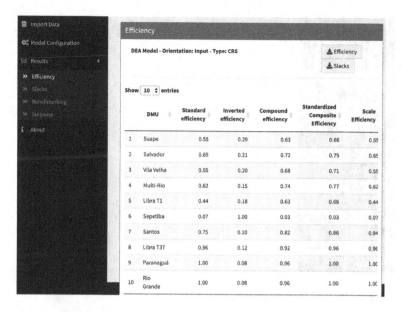

FIGURE 8.6 Efficiency analysis.

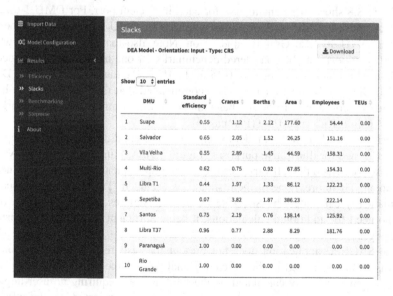

FIGURE 8.7 Analysis of slack.

54 employees redundant. This slack is the guideline managers can rely on for the establishment of improvement plans. Another point that can be observed is that the ports of Paranaguá and Rio Grande do not have slack. This is because these ports were considered to be 100% efficient (Eff. Default=1). In this case, these DMUs (ports) are using the resources in an optimal way and serve as benchmarks for the other ports.

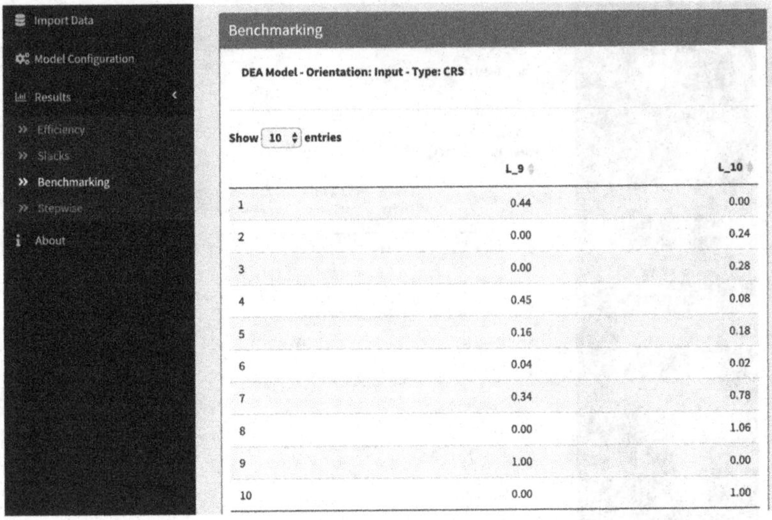

FIGURE 8.8   Analysis of benchmarks.

Figure 8.8 shows the benchmarks for each inefficient port. For DMU 1 (Suape) the benchmark is DMU 9 (Paranaguá), whereas, for DMU 2 (Salvador), the benchmark is DMU 10 (Rio Grande), and so on. Note also that there are cases where two efficient ports can be considered benchmarks for one inefficient DMU, such as DMU 4 (Multi Rio) with values of 0.45 for DMU 9 (Paranaguá) and 0.08 for DMU 10 (Rio Grande). In such a case, it is recommended that the DMU with the highest value be considered to be the benchmark. In this example, the benchmark for DMU 4 (Multi Rio) is DMU 9 (Paranaguá).

Finally, even though the model example showed good discrimination, Stepwise was performed for illustrative purposes (Figure 8.9). As can be seen, the analysis of Stepwise confirmed that all variables used were important and should be retained in the model. Therefore, none of them should be excluded. The SAGEPE application automatically calculates the Stepwise for all the analyses performed. Thus, there was no need for additional calculations, it being sufficient for the user to perform the analysis.

The SAGEPE application does not present the resource of restrictions on the weights to be calculated in the efficiency analyses. In the vast majority of cases, these weights are freely calculated by the DEA model, requiring no user-imposed restrictions. Due to its low usage, this feature was not incorporated into the software. We prefer to adhere to more commonly used aspects, such as the Stepwise procedure, which, if calculated manually, becomes very laborious.

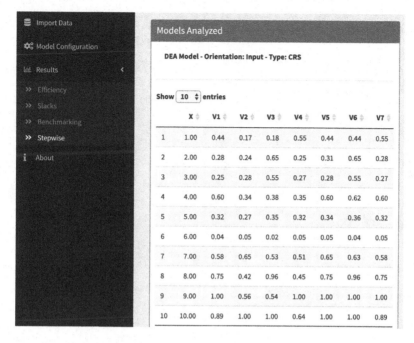

FIGURE 8.9 Stepwise analysis.

FIGURE

9 Analysis in Two Stages: Techniques for Joint Assessment with DEA

This chapter describes the concepts of five techniques that can be used to complement the understanding of efficiency after analysis, using DEA. This procedure is known as two-stage analysis. The techniques are Bootstrapping, CausalImpact, Tobit Regression and Ordinal Least Squares (OLS), Malmquist Index and Artificial Neural Networks (ANNs). In this chapter we will give a brief description of each technique. We strongly recommend that these be used in conjunction with DEA.

There are recurring cases in which researchers perform two procedures in productivity and efficiency analyses. The first procedure is to perform a productivity and efficiency analysis using the DEA technique. The second is performed through statistical interpretations of the estimates of the efficiency scores, as well as the use of regression techniques of efficiency scores in explanatory variables. Both procedures are important for decision-makers to use the DEA analysis.

DEA does not incorporate a statistical concept in its original methodology. Thus, there is a particular research flow that integrates two statistical methodologies in DEA. The first is Bootstrapping, that constructs a basis for statistical inference in DEA. The second establishes procedures for analysis of contextual factors. Finding the contextual factors that affect efficiency is necessary in many DEA applications and is emphasized in many studies (Liu et al., 2016). For these analyses, techniques such as CausalImpact, Tobit Regression and OLS, Malmquist Index and RNA, among others, can be used.

9.1 BOOTSTRAPPING

In order to correct the efficiency values in view of the random error inherent in the data used, Simar and Wilson (1998) proposed a bootstrap analytical approach, through statistical inference, the efficiency results obtained through DEA.

Bootstrapping refers to a repeated regeneration of the original input and output data, according to a specified statistical distribution. The aim of the technique is to imitate the sampling distribution of the original data, allowing estimations of bias in these data. Besides this, confidence intervals are constructed for the efficiency scores (Liu et al., 2016). Thus, one of the objectives of Bootstrapping is to control the sample variance (Førsund, 2017).

Bootstrapping is a statistical technique based on generating a large number of pseudo-observations, formed by the data-generating process assumed (Simar & Wilson, 2000). The sampling bias can then be estimated, and confidence intervals established, introducing a measure of how reliable the efficiency scores are (Førsund, 2017).

Thus, for each unit analyzed, it is possible to estimate the confidence interval of the efficiency, bias and corrected efficiency, which will be considered for the assessment performed using DEA (Périco et al., 2017). In addition to making the scores obtained by calculating the efficiency of DEA more reliable, that is, with fewer problems related to the randomness of the data, Bootstrapping can also contribute to solving problems of discrimination.

To show the application of the Bootstrapping technique in the DEA efficiency scores, we have created an illustrative example (Chart 9.1). An assessment of efficiency can be considered in the production system of five different companies (Companies A to E). The efficiency represents the calculation obtained with DEA. The bias indicates the random error in each efficiency score. Thus, it is possible to calculate the corrected efficiency by subtracting the bias value from the efficiency value.

It can be observed that, as well as correcting the efficiency scores, the Bootstrapping application also changed the ranking of the DMUs, when the DMUs are classified from the most efficient to the least efficient ones. These adjustments can sometimes change the benchmarking performed in the DEA analyses. This is because the DMU considered to be the benchmark in the DEA efficiency analysis may not be the benchmark in the corrected DEA efficiency analysis. This situation is illustrated in the example in Chart 9.1, where, in terms of DEA efficiency, DMU A is considered to present the best performance (benchmark). However, in terms of the corrected efficiency, the best-performing DMU is B.

9.2   CAUSAL IMPACT FOR CAUSAL ANALYSIS

Causal modeling provides the ability to combine cause-and-effect information with statistical data to provide a quantitative assessment of the relationships among the variables studied (Anderson & Vastag, 2004). In the area of operations and service management, the use of causal modeling may help with analyzing the relationships among the variables and provide a basis for understanding the effects of an intervention. The description, assessment and summary of the causal relationships assumed

CHART 9.1

Efficiency with Bootstrapping Application

Company	Efficiency	Bias	Corrected Efficiency	DMU Ranking: Efficiency	DMU Ranking: Corrected Efficiency
A	1.00	0.20	0.80	1	2
B	0.90	0.05	0.85	2	1
C	0.80	0.05	0.75	3	3
D	0.75	0.15	0.60	4	5
E	0.70	0.05	0.65	5	4

are the explanatory components. These relationships can be used to develop diagnoses of the inferences regarding their causes and effects, and the prediction of results that would follow from an intervention, or not (Brodersen et al., 2015).

The use of causality analysis in operations and service management is important because it is possible to verify, for example, whether or not the implementation of an improvement program has positive effects; in other words, whether there is a causality among improvement programs and an increase in overall operating performance (Anderson & Vastag, 2004; Iyer et al., 2013; Brodersen et al., 2015).

However, assessing the effects of improvement programs on efficiency can be considered to be a critical process. This is because there may be a series of other factors, besides the improvement programs implemented, that influence the effects on the efficiency of an organization or a production system (Iyer et al., 2013). Among other factors, one could cite automation, prioritization of a project or product, and changes in management. Often, positive effects are pointed out indicating the improvement of a process without effective empirical verification of causality between the improvement and the effect generated (Boysen & Bock, 2011). Thus, there arises the challenge of isolating and proving the observed effect. Usually, this type of problem is called causal inference analysis in a time series.

A time series can be understood as a sequence of observations of a variable made over time (Ausín et al., 2014). A causal inference can be understood to be an intervention or a treatment suffered in a time series (Brodersen et al., 2015). Thus, the analysis of causal inference in a time series is a way of measuring the effect of a cause (Antonakis et al., 2010).

Brodersen et al. (2015) discussed the importance of assessing the effect of a causal inference on a time series, such as assessing the effect, among others, of: (i) introducing a new product on to the market; (ii) an advertising campaign; or (iii) improvement programs implemented in companies. Seeking to assist managers and researchers with such assessments, the research by Brodersen et al., (2015) presented the CausalImpact technique.

The CausalImpact technique was developed by researchers and professionals at Google to find a way to measure the causal effect of a market intervention. Brodersen et al. (2015) illustrated the application of the technique by analyzing the effect of an advertising campaign by one of Google's U.S. advertisers. The survey studied the causal effects on the number of times an Internet user was directed to the advertiser's site from the Google search page and estimated how the number of clicks on the same advertiser's pages would have changed if the advertising campaign had not been developed. In this manner, the effects of one's advertisements can be measured.

According to Brodersen et al. (2015), the technique, developed to calculate the effect of causal inference, produces the modeling of inferred responses of an observed time series. This is achieved by considering the periods before and after the intervention, the technique being based on two aspects: (i) providing a time series with the estimated effect based on the Bayesian series, and (ii) using the average model to construct the most appropriate synthetic control to model the inference.

The inference can be understood as what would have happened to the outcome of the time series in the absence of the treatment (Brodersen et al., 2015). In the Bayesian approach, the current knowledge about the model parameters is expressed in terms of a probability distribution called an a priori distribution. When new data containing information related to model parameters become available, it is understood that a probability proportional to the distribution of the observed data can be expressed. This information is then combined with the a priori distribution, producing an updated probability distribution called the a posteriori distribution, on which the Bayesian inference is based (Ausin et al., 2014).

According to Brodersen et al. (2015), in general terms, there are three sources of information available for the construction of adequate synthetic control. The first is the behavior of the response time series itself before the intervention. The second is the behavior of other time series that had not undergone the intervention (control time series). The third source of information is prior knowledge available about the model parameters.

It is understood that, although it was developed for application in advertising campaigns, the CausalImpact technique can be used in several other applications. CausalImpact can be calculated using R software.

9.3   TOBIT REGRESSION AND ORDINAL LEAST SQUARES (OLS)

The DEA technique and regression analysis are alternative methods that can be used together to compare efficiency in organizations. Linear regressions can be carried out as simple or multiple regressions. Simple linear regression allows analysis of the relationship between a single dependent variable and a single independent variable. Multiple linear regression allows analysis of the relationship between a single dependent variable and two or more independent variables. For both cases, the validation of the null hypothesis h0 and the alternative hypothesis h1 depends on the P value obtained in the test (Corrar et al., 2007). Dancey and Reidy (2013) highlight the main analyses in a linear regression:

i)   correlation between x and y, represented by the Pearson R; in regression analysis, it is also known as multiple R, which indicates the degree of correlation between the variables. This correlation indicates the grouping proximity of the points around the line of best fit;

ii) variance explained, represented by R^2. The correlation coefficient is squared to obtain a measure of variance explained by the relationship;

iii) adjusted R^2: provides a more realistic estimate, as it considers the population data, rather than the sample data;

iv) standard error: an estimate of the variance of y for each value of x, providing information on how correct the estimate can be.

Unlike simple and multiple linear regressions, the Tobit regression was developed with the objective of predicting models with limited dependent variables. The basis of the Tobit model is similar to linear regression, but it assumes a truncated or censored normal distribution, and becomes an effective method to estimate the relationship between a truncated or censored dependent variable and other independent variables (Amemiya, 1984).

A truncated or censored distribution is a subset of a distribution with data above or below a given value. For example, for research on individuals who have an income between certain values, the sample would be limited (truncated) between these values (Greene, 1997).

The efficiency analysis models with DEA produce results between 0% and 100% or 0 and 1. However, there is divergence in the literature as to whether this characteristic makes the efficiency scores truncated or fractional (Simar & Wilson, 2007; McDonald, 2009). Thus, there are some authors (Simar & Wilson, 2007, 2011) who defend the use of the Tobit regression, while others (Banker & Natarajan, 2008; McDonald, 2009) defend regression based on ordinal least squares. We, the authors of this book, recommend using the Tobit regression.

To perform the Tobit regression, it is necessary to fulfill a set of assumptions, such as normality of residues, homoscedasticity of residues, absence of serial autocorrelation in residues, and multicollinearity among independent variables (Corrar et al., 2007).

The assumption of normality indicates whether the residues generated in the entire range of observations come from a normal distribution. The homoscedasticity indicates whether the residuals for each observation of X have constant variance over the full extent of the independent variables. The absence of serial autocorrelation indicates that the residue should be independent between X_t and $X_t - 1$; values close to 2 satisfy this assumption. The multicollinearity between the independent variables occurs when two or more independent variables contain information similar to the dependent variable; values from 1 to 10 meet this assumption (Corrar et al., 2007). Tobit regression tests can be run in software R, version 0.1, package "censReg".

9.4 MALMQUIST INDEX

This index was created by Malmquist (1953), and reformulated by Caves et al. (1982). It is used to assess the relative evolution of efficiency and productivity of each DMU of a sample being studied. When there are time series data, also called panel data, in more than one analysis unit, one can calculate the necessary distance measurements for the Malmquist Index using the DEA technique (Färe et al., 1994).

The Malmquist Index assesses the productivity change of a DMU between two time periods (t and t + 1) as a product of the term of recovery and the change of frontier. The term "recovery" reflects the degree by which a DMU reaches for the efficiency to be improved, while the term "frontier change" (or innovation) reflects the change in the efficient frontier around the DMU between the two periods (Örkcü et al., 2016).

The Malmquist Index provides the DMU Total Factor Productivity Index (TFPI), as it reflects the progress or regression in the DMU efficiency along with the progress or regression of the frontier technology. Thus, the total factor productivity representing overall productivity can be broken down into two mutually exclusive components: (i) one responsible for measuring a change in technical efficiency (convergence effect); and (ii) another measuring a change in technology (frontier change) (Örkcü et al., 2016).

The Total Factor Productivity Index (TFPI) is defined as the product of two other indexes. The first is the Technical Efficiency Change Index (TECI), and the second, the Technological Change Index (TCI) (Ahn & Min, 2014), according to Equation 9.1:

$$TFPI = TECI \times TCI \qquad (9.1)$$

The Technical Efficiency Change Index (TECI) is related to the degree of effort made by the DMU to improve its efficiency. Basically, it measures the capacity of the analysis unit to approach or move away from frontier technology from t to t + 1. The technological change index (TCI) reflects the change in the efficient frontiers surrounding the DMUs between the two periods (t and t + 1) and measures the change in the output set or the movement of the production frontier between the periods (Ahn & Min, 2014; Aparicio et al., 2016).

The Malmquist Index provides assistance for analysis of productivity and efficiency, as it allows identification of whether there was technological progress or improvement in technical efficiency, or both. Therefore, these distances may present results lower, equal to or greater than 1. In addition, it is important to distinguish between productivity changes due to technical change and technological change. This is because, with this distinction, it is possible to observe when productivity gains that generate frontier displacement are provided by innovations inserted in the process or by the system under analysis. In addition, it is also possible to see whether productivity gains could be a consequence of the diffusion of technology or the process of learning and continuous improvement.

Thus, the distinction among results becomes very important for decision-making, because, when there is no technical progress, it is fundamental to increase the research process and to implement projects to raise the efficiency frontier. Moreover, when the evolution is being delayed by displacement of the units in relation to the frontier, there are possible problems in the diffusion and use of technological innovations. The joint DEA analysis with the Malmquist Index can be performed by the Data Envelopment Analysis Application (DEAP) application.

9.5 ARTIFICIAL NEURAL NETWORKS (ANNS)

Artificial neural networks (ANNs) are part of an artificial intelligence (AI) field that seeks to develop intelligent programs employing models that imitate the functional structure of neurons in the human brain. More specifically, the objective of AI is to develop methodologies or computational devices that possess or multiply rational human ability to reason, perceive, make decisions and solve problems (Luger, 2014).

The first applications of ANNs occurred in the 1940s, but the research intensified and underwent increased applicability in the 1990s. This fact is due mainly to the scenario of favorable conditions that were established, such as the technological evolution that enabled the increase in capacity of computer processing. In addition,

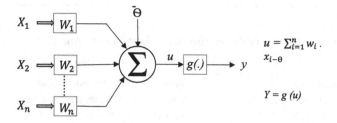

FIGURE 9.1   Mathematical model of the artificial neuron. (Adapted from Silva et al., 2010.)

advances occurred in the elaboration of optimization algorithms and in the study of the biological nervous system. The practical applications of ANNs are present in several situations in our daily lives, such as control of machines and equipment, medical diagnoses, identification of credit profiles and forecast of actions in the financial market, among others (Silva et al., 2010).

In general, an ANN can be described as a data-processing model implemented by artificial neurons that communicate through innumerable connections (artificial synapses), and are capable of acquiring and maintaining knowledge based on information. Furthermore, they are able to recognize patterns. ANNs imitate the human brain in two respects: (i) the ability to learn through information adapted to context; and (ii) the ability to store knowledge through synaptic forces, represented by synaptic weights (Marques et al., 2014).

The artificial neuron is the fundamental information processing unit in an ANN. The functioning of an artificial neuron can be described as follows: (i) determination of the input variables; (ii) multiplication of the data of each input by its corresponding synaptic weight; (iii) calculation of the activation potential obtained by the weighted sum of the input signals minus the activation threshold; (iv) use of the appropriate activation function that restricts the output neuron; and (v) calculation of the output result (Silva et al., 2010). ANNs are composed of neurons, which receive input data, process and transfer the results through a single output, which can be the final output of the network or pass the information onto another neuron (Haykin, 2001). Figure 9.1 and Chart 9.2 show an example of a mathematical model of an artificial neuron.

The ANNs are also characterized by overall properties, such as the network topology, the learning algorithm used and the coding scheme (Haykin, 2001). The network topology refers to the connection pattern among the neurons. The learning algorithm and the coding scheme are equivalent to the interpretation of the data provided to the network in relation to the result of their processing (Luger, 2014).

In order to obtain consistent results from the ANNs, it is necessary that some procedures be observed, such as: (i) data preparation; (ii) network construction; (iii) training; (iv) testing; and (v) interpretation of results. Moreover, it is critical to clearly define the problem to be addressed by the ANN prior to commencing its construction.

CHART 9.2

Description of the Variables of the Artificial Neuron Model

Element	Description
Input signals: $\{x1, x2, ..., xn\}$	Values assumed by the input variables.
Synaptic weights: $\{w1, w2, ..., wn\}$	Values for weighting the input variables.
Linear combiner: $\{\Sigma\}$	Aggregating function, composed of the sum of all the input signals multiplied by their respective weights.
Activation threshold: $\{\theta\}$	A variable that determines which level will be appropriate so that the result produced by the linear combiner generates activation of the neuron.
Activation potential: $\{u\}$	$$u = \sum_{i=1}^{n} w_i \cdot x_{i-\theta}$$ If $u \geq \theta$, the neuron produces a potential that activates the neuron; otherwise, the potential will prevent activation.
Activation function: $\{g.\}$	Limits the output neuron within the interval of allowed values.
Output signal: $\{y\}$	Final result produced by the neuron based on the set of input signals, which can still be used as input signals for other neurons.

Source: Adapted from Silva et al. (2010).

REFERENCES

Ahn, Y. H., & Min, H. (2014). Evaluating the multi-period operating efficiency of international airports using data envelopment analysis and the Malmquist productivity index. *Journal of Air Transport Management*, *39*(2), 12–22.

Amemiya, T. (1984). Tobit models: A survey. *Journal of Econometrics*, *24*(12), 3–61.

Anderson, R. D., & Vastag, G. (2004). Causal modeling alternatives in operations research: Overview and application. *European Journal of Operational Research*, *156*(1), 92–109.

Antonakis, J., Bendahan, S., Jacquart, P., & Lalive, R. (2010). On making causal claims: A review and recommendations. *The Leadership Quarterly*, *21*(6), 1086–1120.

Aparicio, J., Garcia-Nove, E. M., Kapelko, M., & Pastor, J. T. (2016, November). Graph productivity change measure using the least distance to the pareto-efficient frontier in data envelopment analysis. *Omega*, *72*(1), 1–14.

Ausín, M. C., Galeano, P., & Ghosh, P. (2014). A semiparametric Bayesian approach to the analysis of financial time series with applications to value at risk estimation. *European Journal of Operational Research*, *232*(2), 350–358.

Banker, R. D., & Natarajan, R. (2008). Evaluating contextual variables affecting productivity using data envelopment analysis. *Operations Research*, *56*(1), 48–5.

Boysen, N., & Bock, S. (2011). Scheduling just-in-time part supply for mixed-model assembly lines. *European Journal of Operational Research*, *211*(1), 15–25.

Brodersen, K. H., Gallusser, F., Koehler, J., Remy, N., & Scott, S. L. (2015). Inferring causal impact using Bayesian structural time-series models. *The Annals of Applied Statistics*, *9*(1), 247–274.

Caves, D. W., Christensen, L. R., & Diewert, W. E. (1982). The economic theory of index numbers and the measurement of input, output, and productivity. *Econometrica: Journal of the Econometric Society, 50*(6), 1393–1414.

Corrar, L. J., Paulo, E., & Dias Filho, J. M. (2007). *Análise multivariada: para os cursos de administração, ciências contábeis e economia* (pp. 280–323). São Paulo: Atlas.

Dancey, C. P., & Reidy, J. (2013). *Estatística sem matemática para psicologia.* Penso Editora.

Färe, R., Lovell, C. K., & Grosskopf, S. (1994). *Production frontiers.* Cambridge University Press.

Førsund, F. R. (2017). Economic interpretations of DEA. *Socio-Economic Planning Sciences, 61*(1), 9–15.

Greene, W. (1997). *Econometric analysis.* Upper Saddle River, NJ: Prentice-Hall International.

Haykin, S., (2001). *Redes neurais: princípios e prática* (2nd ed.). Porto Alegre: Artes Médicas.

Iyer, A., Saranga, H., & Seshadri, S. (2013). Effect of quality management systems and total quality management on productivity before and after: Empirical evidence from the Indian auto component industry. *Production and Operations Management, 22*(2), 283–301.

Liu, J. S., Lu, L. Y., & Lu, W. M. (2016). Research fronts in data envelopment analysis. *Omega, 58*, 33–45.

Luger, G.F., (2014). *Inteligência artificial* (6th ed.). São Paulo: Pearson.

Malmquist, S. (1953). Index numbers and indifference surfaces. *Trabajos de estadística, 4*(2), 209–242.

Marques, A., Lacerda, D. P., Camargo, L. F. R., & Teixeira, R. (2014). Exploring the relationship between marketing and operations: Neural network analysis of marketing decision impacts on delivery performance. *International Journal of Production Economics, 153*, 178–190.

McDonald, J. (2009). Using least squares and Tobit in second stage DEA efficiency analyses. *European Journal of Operational Research, 197*(2), 792–798.

Örkcü, H. H., Balıkçı, C., Dogan, M. I., & Genç, A. (2016). An evaluation of the operational efficiency of Turkish airports using data envelopment analysis and the Malmquist productivity index: 2009–2014 case. *Transport Policy, 48*(1), 92–104.

Périco, A. E., Santana, N. B., & Rebelatto, D. A. D. N. (2017). Efficiency of Brazilian international airports: Applying the bootstrap data envelopment analysis. *Gestão & Produção, 24*(2), 370–381.

Silva, I. N., Spatti, D. H., & Flauzino, R. A., (2010). *Redes neurais artificiais para engenharia e ciências aplicadas.* São Paulo: Artliber.

Simar, L., & Wilson, P. W. (1998). Sensitivity analysis of efficiency scores: How to bootstrap in non-parametric frontier models. *Management Science, 44*(1), 49–61.

Simar, L., & Wilson, P. W. (2000). Statistical inference in non-parametric frontier models: The state of the art. *Journal of Productivity Analysis, 13*(1), 49–78.

Simar, L., & Wilson, P. W. (2007). Estimation and inference in two-stage, semi-parametric models of production processes. *Journal of Econometrics, 136*(1), 31–64.

Simar, L., & Wilson, P. W. (2011). Two-stage DEA: Caveat emptor. *Journal of Productivity Analysis, 36*(2), 205.

10 DEA Application Examples

After explaining DEA in the course of this book, we trust that the reader is now better prepared to understand the results in relation to the practical problems presented in Chapter 3. We have made these explanations in this final part of the book to generate reflection on the application potential of the technique.

10.1 EFFICIENCY AND INTERVENTION IN THE SYSTEM: CASE STUDY IN A BUS MANUFACTURING COMPANY

Initially, we will present the results regarding the practical problem of productivity and efficiency management discussed in Section 3.1. The working method used in developing the research was presented in Chapter 7, Section 7.1. Thus, we will go directly to the results. The analysis will be presented, divided into Product Engineering and the Production Process.

10.1.1 ANALYSIS AND DISCUSSION OF RESULTS OF PRODUCT ENGINEERING

Figure 10.1 illustrates the results regarding composite technical efficiency of the modularized product project (MPP) and the non-modularized product project (nMPP). This analysis makes it possible to visualize the evolution of efficiency scores over time in Product Engineering and is segregated (black dashed line) in the transition period, the period in which modularization was implemented in the Company's Product Engineering.

By analyzing Figure 10.1, it is possible to observe an increase in the efficiency of the MPP over time, noting that the DMUs with the best efficiency performance scores are located in the period after modularization. It is also observed that, after the rise in the efficiency scores in the period after modularization, there was a reduction in the DMU scores, namely, 39 (March 2014) and 40 (April 2014), which was related to a design problem in the doors and space dividers of the product, which led to an increase in client complaints (input 10). However, it is possible to visualize different effects in relation to the efficiency of the nMPP, considering that the DMUs with the best performance in the time sequence are concentrated in the period prior to modularization (2011). Chart 10.1 shows the composite technical efficiency scores shown in Figure 10.1.

The features of Product Engineering are shared, with any improvement action performed by Engineering (automation, for example) affecting both types of projects (MPP and nMPP) evenly. Following a discussion with the process specialists, it was observed that there was no action that prioritized MPP to the detriment of nMPP

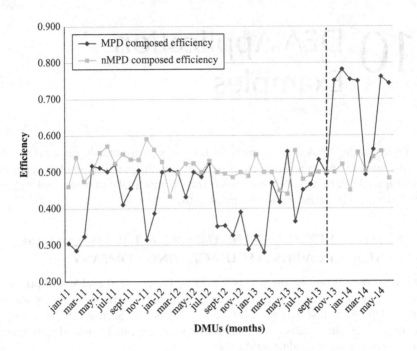

FIGURE 10.1   Evolution of efficiency of modularized and non-modularized product project.

(prioritization of one product project over another, for example). Chart 10.2 summarizes the means of efficiency scores for product designs, of the periods before and after modularization, including the overall mean efficiency.

When analyzing Chart 10.2, it can be observed that the MPP resulted in an increase in mean efficiency after modularization (before: 0.428; after: 0.700). The minimum efficiency of MPP occurred in the period prior to modularization (February 2013) and the maximum efficiency score occurred in the period after modularization (December 2013). It is understood that the location of the maximum efficiency score in the period after modularization of the product design is indicative of the positive effects provided by modularization. The efficiency of the nMPP did not exhibit the same behavior, showing that a significant increase in efficiency between the periods before and after modularization was not observed. Regarding the periods of minimum efficiency (0.433) and maximum efficiency (0.591) of nMPP, both were located in the period prior to modularization, in February 2012 and November 2011, respectively.

Based on the data presented, it is understood that the analyses of the technical efficiencies associated with the MPP and nMPP arranged in time series showed indications that the modularization of products provided improvements in the efficiency of Product Engineering. However, to confirm that the mean efficiency scores comparing the periods before and after modularization were significantly different, statistical tests (ANOVA and assumptions) were performed. In order to determine whether there was causality between the effects perceived in the efficiency and the modularization of products, the causal inference analysis (CausalImpact) was carried out.

CHART 10.1
Technical Efficiency of the Modularized and Non-Modularized Product Project

DMU	Month/Year	Composite technical efficiency of the MPP	Composite technical efficiency of the nMPP
DMU1	Jan/11	0.305	0.461
DMU2	Feb/11	0.285	0.541
DMU3	Mar/11	0.324	0.475
DMU4	Apr/11	0.518	0.500
DMU5	May/11	0.512	0.553
DMU6	Jun/11	0.501	0.571
DMU7	Jul/11	0.522	0.524
DMU8	Aug/11	0.410	0.550
DMU9	Sep/11	0.455	0.535
DMU10	Oct/11	0.505	0.534
DMU11	Nov/11	0.314	0.591
DMU12	Dec/11	0.386	0.561
DMU13	Jan/12	0.500	0.529
DMU14	Feb/12	0.506	0.433
DMU15	Mar/12	0.500	0.497
DMU16	Apr/12	0.432	0.523
DMU17	May/12	0.500	0.524
DMU18	Jun/12	0.487	0.500
DMU19	Jul/12	0.524	0.531
DMU20	Aug/12	0.351	0.500
DMU21	Sep/12	0.353	0.496
DMU22	Oct/12	0.326	0.486
DMU23	Nov/12	0.390	0.500
DMU24	Dec/12	0.288	0.489
DMU25	Jan/13	0.324	0.548
DMU26	Feb/13	0.279	0.500
DMU27	Mar/13	0.470	0.500
DMU28	Apr/13	0.417	0.449
DMU29	May/13	0.556	0.440
DMU30	Jun/13	0.364	0.559
DMU31	Jul/13	0.451	0.481
DMU32	Aug/13	0.466	0.492
DMU33	Sep/13	0.534	0.500
DMU34	Oct/13	0.494	0.500
DMU35	Nov/13	0.751	0.500
DMU36	Dec/13	0.782	0.521
DMU37	Jan/14	0.754	0.450
DMU38	Feb/14	0.750	0.553
DMU39	Mar/14	0.492	0.504
DMU40	Apr/14	0.561	0.541
DMU41	May/14	0.761	0.557
DMU42	Jun/14	0.743	0.483

Legend:

Period prior to modularization
Period of transition to modularization
Period after modularization

CHART 10.2

Analysis of the Efficiency of Groups Before and After Modularization (Product Engineering)

	Modularized Product Project (MPP)			Non-Modularized Product Project (nMPP)		
	Before modularization	After modularization	General	Before modularization	After modularization	General
Mean	0.428	0.700	0.480	0.511	0.513	0.512
Standard Deviation	0.087	0.109	0.140	0.037	0.037	0.037
Minimum	0.279	0.492	0.279	0.433	0.450	0.433
Maximum	0.556	0.782	0.782	0.591	0.557	0.591

10.1.2 PRESUMPTIONS AND ANALYSIS OF VARIANCE (ANOVA) OF PRODUCT ENGINEERING EFFICIENCY

To perform the ANOVA test, it was first necessary to fulfill some assumptions, which are data normality (Shapiro-Wilk test) and data variance homogeneity (Levene's test). With respect to the modularized product project, the Shapiro-Wilk test of the periods before and after modularization revealed a significance level greater than 0.05 (P before = 0.639 and P after = 0.220). As for the Levene's test, the result obtained was $P = 0.566$. In this way, it was verified that the data approximated to a normal distribution and were homogeneous. With respect to the non-modularized product project, the Shapiro-Wilk test (P before = 0.596 and P after = 0.703) and the Levene's test ($P= 0.958$) revealed that the data also approximated to a normal distribution and were homogeneous. In this way, the assumptions for the use of ANOVA were fulfilled, and the ANOVA statistical test could be carried out (Chart 10.3).

The results of the calculation of the ANOVA provided evidence that, in the modularized product project. the mean of the composite technical efficiency of Product Engineering increased from 0.428 before to 0.700 after modularization. The F score, 57.688, of the means of the periods considered in the modularized product project,

CHART 10.3

Analysis of the Efficiency of the Groups Before and After Modularization in the Product Engineering

Period	Mean Efficiency of the MPP	Mean Efficiency of the nMPP
Before modularization	0.428	0.511
After modularization	0.700	0.513
Difference in efficiency (amplitude)	0.272	0.002
F	57.688	0.032
P-value	0.000	0.858

and the *P*-value of 0.000, confirm that the difference detected was statistically significant.

As for the non-modularized product project, the same assessments were carried out for confirmatory purposes. It can be noticed that the efficiency in the period prior to modularization was 0.511 and in the subsequent period, 0.513. However, the F of 0.032 and the *P*-value, 0.858, indicated that that there was no significant difference between the mean efficiencies of the two periods.

Based on the analyses carried out, it is understood that there are effects of the modularization of products on the efficiency of the Product Engineering. These effects are positive, with modularization resulting in an increase in efficiency. It is understood that this finding shows that the improvement in efficiency over time observed in Product Engineering can be attributed to the implementation of modularization.

10.1.3 CAUSALITY AND DIMENSIONING OF THE EFFECTS OF MODULARIZATION OF PRODUCTS ON THE EFFICIENCY OF THE PRODUCT ENGINEERING

With the data envelopment analysis technique, it was possible to verify the effect of modularization on the efficiency of the company's Product Engineering. However, with the use of DEA, it is not possible to accurately measure the magnitude of the observed effect, since the behavior of the efficiency of the modularized product project is not known for the period from November 2013 to July 2014 (representing the period before modularization, i.e. when the company had still not implemented it). In order to carry out this assessment, the causal impact caused by the modularization with the use of CausalImpact was analyzed.

With the control variable (nMPP efficiency scores) and the response variable (MPP efficiency scores), and also with the history of the response variable in the period prior to treatment, the CausalImpact technique performs statistical calculations that include the absolute mean effect and the relative effect caused by the intervention (Brodersen et al., 2015). Figure 10.2 illustrates the behavior of the modularized product project efficiency over time and the confractual estimate, taking into account the 42 DMUs analyzed. The dotted line between DMUs 30 and 40

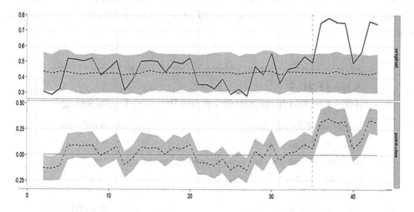

FIGURE 10.2 Measuring the effect of modularization on the Product Engineering.

represents the period in which the series was modularized. The shaded line represents the efficiency variations over time, considering the period before modularization. The dotted line represents the average behavior of the efficiency series prior to modularization and estimates what the behavior would have been if the modularization had not been implemented.

In analyzing Figure 10.2, it is necessary to consider that the "original" behavior contemplates the current behavior of the project efficiency of the modularized product and the behavioral projection if the project had not been modularized. The current behavior is illustrated by the black line, reflecting the same behavior as the efficiency shown previously in Figure 10.1. Thus, when analyzing the "original" behavior of the time series in the period after modularization, it should be considered that the behavior of the current efficiency (black line), which is above the shaded part, is the efficiency increase obtained by the implementation of modularization in Product Engineering. If the modularization had not been implemented, the variation of efficiency in the period after modularization would have been restricted to the variations within the shaded line and not above, as observed.

Regarding the behavior of the efficiency series in Figure 10.2, as "pointwise" the CausalImpact technique estimates the behavior of the effect in a 95% confidence interval. Thus, the shaded strip shows how the variation within this interval would be. Chart 10.4 summarizes the amplitude of the effects of modularization in Product Engineering.

Upon analysis of the results presented in Chart 10.4, we can observe the current scenario, where the mean composite technical efficiency in the period after modularization in Product Engineering is 0.700. It is also possible to observe the comparative scenario in the post-modularization period; if the modularization had not been implemented, the mean technical efficiency of the period considered after modularization would have been 0.430. Therefore, it can be concluded that, with the implementation of modularization, there was an increase of 0.270 in the mean composite technical efficiency of Product Engineering. It is also possible to observe the relative effect, which shows an increase of 63%. It should be noted that the probability of the causal effect was 99.88901%, greater than the 95% stipulated as a premise for the model. In view of the above, it has been confirmed that there is causality between the modularization of products and the efficiency of Product Engineering.

CHART 10.4

Effect of Modularization on the Product Engineering

Behavior of the efficiency	Mean efficiency
Current (modularized)	0.700
Comparative scenario (in case of no modularization)	0.430
Absolute effect	0.270
Relative effect	63%
Probability of causal effect	99.88901

10.1.4 ANALYSIS AND DISCUSSION OF THE RESULTS OF THE PRODUCTION PROCESS

In the analysis regarding the Production Process, the same procedures presented previously were carried out as for Product Engineering. First, there was determination of the efficiency behavior of the Production Process, of the modularized and non-modularized products, as shown in Figure 10.3. The analysis makes it possible to visualize the evolution of the efficiency scores over time, this being segregated to the transition period (black dashed line), the period in which the product changed from integral to modular.

As shown in Figure 10.3, when the period prior to modularization is considered, minor variations are observed from January 2011 to July 2012. There was a reduction in the performance of the DMUs in the second half of 2012 and an increase in 2013. However, with respect to the efficiency of the modularized product, it is observed that the best performance scores are located in the period following modularization (November 2013 to June 2014). With regard to the non-modularized product, it can be seen that the highest efficiency scores were obtained in the DMUs of the period prior to modularization, namely in the months of 2011. Chart 10.5 shows the composite technical efficiency scores shown in Figure 10.3.

The resources of the Production Process are shared, that is, any improvement actions carried out in the operations affected both products (MP and nMP) uniformly and at the same time. Based on a discussion with the process specialists, it was observed that there was no action that prioritized the MP to the detriment of

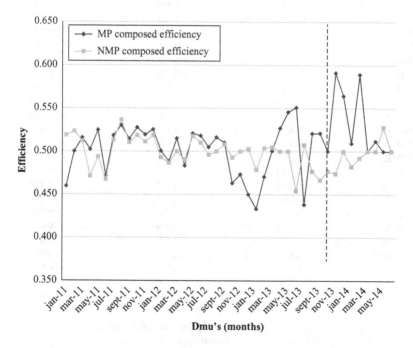

FIGURE 10.3 Evolution of the modularized and non-modularized product efficiency.

CHART 10.5
Efficiencies of the Modularized (MP) and Non-Modularized (nMP) Product

DMU	Month/Year	Composite technical efficiency of the MP	Composite technical efficiency of the nMP
DMU1	Jan/11	0.459	0.519
DMU2	Feb/11	0.500	0.523
DMU3	Mar/11	0.516	0.512
DMU4	Apr/11	0.502	0.471
DMU5	Mai/11	0.525	0.494
DMU6	Jun/11	0.472	0.467
DMU7	Jul/11	0.518	0.513
DMU8	Aug/11	0.530	0.537
DMU9	Sep/11	0.514	0.510
DMU10	Oct/11	0.528	0.518
DMU11	Nov/11	0.519	0.511
DMU12	Dec/11	0.525	0.519
DMU13	Jan/12	0.501	0.493
DMU14	Feb/12	0.488	0.486
DMU15	Mar/12	0.515	0.500
DMU16	Apr/12	0.483	0.490
DMU17	May/12	0.521	0.518
DMU18	Jun/12	0.518	0.510
DMU19	Jul/12	0.505	0.496
DMU20	Aug/12	0.516	0.500
DMU21	Sep/12	0.510	0.508
DMU22	Oct/12	0.463	0.493
DMU23	Nov/12	0.473	0.500
DMU24	Dec/12	0.450	0.503
DMU25	Jan/13	0.433	0.479
DMU26	Feb/13	0.470	0.504
DMU27	Mar/13	0.501	0.505
DMU28	Apr/13	0.527	0.500
DMU29	May/13	0.546	0.500
DMU30	Jun/13	0.551	0.454
DMU31	Jul/13	0.438	0.508
DMU32	Aug/13	0.521	0.477
DMU33	Sep/13	0.521	0.467
DMU34	Oct/13	0.500	0.477
DMU35	Nov/13	0.591	0.474
DMU36	Dec/13	0.564	0.500
DMU37	Jan/14	0.509	0.482
DMU38	Feb/14	0.589	0.492
DMU39	Mar/14	0.500	0.500
DMU40	Apr/14	0.512	0.500
DMU41	May/14	0.500	0.528
DMU42	Jun/14	0.500	0.500

Legend:

Period prior to modularization
Transition period
Period after modularization

the nMP. Chart 10.6 summarizes the product efficiency score means, including the periods before and after modularization, in addition to the overall mean efficiency.

When analyzing Chart 10.6, it can be observed that the MP exhibited an increase in mean efficiency after modularization (before: 0.501; after: 0.533). The minimum efficiency of MP (0.433) was located in the period prior to modularization (January 2013) and the maximum efficiency score (0.591) in the period after modularization (November 2013). It is understood that the location of the maximum efficiency score in the period after product modularization is indicative of the positive effects provided by modularization. The efficiency of the nMP did not exhibit the same behavior, showing that no increase in efficiency was observed between the periods before and after modularization. Regarding the minimum (0.454) and maximum (0.537) efficiency of the nMP, both were located in the period prior to modularization, representing June 2012 and August 2011, respectively.

Based on the data presented, it is considered that the analyses of the technical efficiencies associated with the MP and nMP, arranged in a time series, present indications that suggest that improvements in the efficiency of the Production Process resulted from modularization. However, to confirm whether the means of the efficiency scores considering the periods before and after the modularization were statistically significant, statistical tests (ANOVA and assumptions) were performed. In order to determine whether there is causality between the effect perceived in the efficiency and the modularization of products, causal inference analysis (CausalImpact) was carried out.

10.1.5 ASSUMPTIONS AND ANALYSIS OF VARIANCE (ANOVA) OF THE PRODUCTION PROCESS EFFICIENCY

As for the modularized product, the Shapiro-Wilk test comparing the period before and after modularization showed a significance level greater than 0.05 (prior: 0.212; after: 0.110). As for Levene's test, the result obtained was $P = 0.055$. Thus, it was verified that the data approximated to a normal distribution and were homogeneous.

CHART 10.6

Analysis of the Efficiency of Groups Before and After Modularization (Production Process)

	Modularized product (MP)			Non-Modularized product (nMP)		
	Before modularization	After modularization	General	Before	After	General
Mean	0.501	0.533	0.508	0.499	0.497	0.498
Standard Deviation	0.029	0.041	0.034	0.018	0.016	0.018
Minimum	0.433	0.500	0.433	0.454	0.474	0.454
Maximum	0.551	0.591	0.591	0.537	0.528	0.537

With respect to the non-modularized product, the Shapiro-Wilk test (prior to modularization: $P = 0.375$; subsequent to modularization: $P = 0.234$) and Levene's test ($P = 0.434$), showed that the variables also approximated to a normal distribution and were homogeneous. Thus, the assumptions for the use of ANOVA to compare the means of composite technical efficiencies in the modularized and non-modularized product were met, and the ANOVA test was performed (Chart 10.7).

The results of the ANOVA show that, in the case of the modularized product, the mean of the composite technical efficiency of the Production Process increased from 0.501 to 0.533 after modularization. The F score of the means of the periods considered in the modularized product was 6.425552 and the P value, 0.015264, confirming that the difference detected was statistically significant.

As for the non-modularized product, the same assessments were carried out for confirmatory purposes. It was observed that the efficiency in the period prior to modularization was 0.499, and in the subsequent post-modularization period, 0.497. However, the F value of 0.063992 and the P value of 0.80159 indicated that there was no significant difference between the efficiency means of the two periods. Thus, the efficiency of the non-modularized product did not show significant variation during the periods before and after modularization. This finding indicated that the improvement in efficiency over time observed in the Production Process can be attributed to the implementation of modularization.

Therefore, one can observe that there were effects of the modularization of products on the efficiency of the Production Process. As shown, these effects are positive, in that modularization resulted in increased efficiency.

10.1.6 CAUSALITY AND DIMENSIONING OF THE EFFECTS OF PRODUCT MODULARIZATION ON PRODUCTION PROCESS EFFICIENCY

In order to analyze the effects of product modularization in the Production Process, the CausalImpact technique was used. Figure 10.4 illustrates the efficiency behavior of the modularized product over time.

Upon analysis of the "original" behavior of the time series, based on estimation of the period after modularization, it should be considered that the current efficiency

---

CHART 10.7
ANOVA Test of the Production Process

Period	Mean of the composite technical efficiency of the Modularized Product (MP)	Mean of the composite technical efficiency of the Non-Modularized Product (nMP)
Before modularization	0.501	0.499
After modularization	0.533	0.497
Difference in efficiency (amplitude)	0.032	−0.002
F	6.425552	0.063992
p-value	0.015264	0.80159

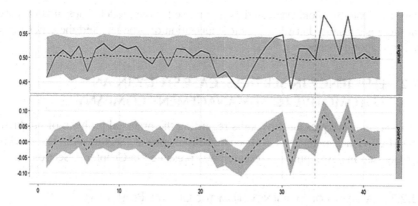

FIGURE 10.4   Effect of modularization on the Production Process.

behavior, marked by the black line above the shaded area, is the efficiency gain obtained with the implementation of modularization. If the modularization had not been implemented, the efficiency variation would have remained within the shaded line, and not above, as observed. As for the "pointwise", an effect was estimated as a 95% confidence interval. Thus, the shaded line shows how the upward and downward variation would be within this interval. Chart 10.8 summarizes the breadth of the effects of the implementation of product modularization in the Production Process.

In analyzing the results presented in Chart 10.8, we can observe the current scenario, with the mean efficiency of the modularized product in the post-modularization period being 0.533. It is also possible to observe the comparative scenario in the same period, where, if modularization had not been implemented, the mean technical efficiency of the period considered after modularization would have been 0.501. It can be concluded, therefore, that the implementation of product modularization generated an increase of 0.032 in the Production Process efficiency. The results are statistically significant with respect to both the ANOVA and CausalImpact tests. It is also possible to observe the relative effect, which reveals an increase of 6.3%. It is also worth noting that the probability of the causal effect was 99.88901%, well

CHART 10.8
**Effect of Modularization on the Efficiency of the
Production Process**

Behavior of the efficiency	Mean Efficiency
Current scenario (modularized)	0.533
Confractual scenario (if no modularization)	0.501
Absolute effect	0.032
Relative effect	6.38%
Probability of causal effect	99.88901

above the 95% threshold stipulated as a premise for the model. Thus, it is observed that there is causality between the modularization of products and the efficiency of the Production Process.

10.2 EFFICIENCY IN SERVICES: CASE STUDY IN AN AUTOMOTIVE FLEET MANAGEMENT COMPANY

In this section, we present the results regarding the practical problem of productivity and efficiency management discussed in Chapter 3, Section 3.2. The working method used in developing the research was presented in Chapter 7, Section 7.2.

10.2.1 ANALYSIS OF EFFICIENCY FROM THE CLIENT'S PERSPECTIVE

Chart 10.9 presents a consolidated analysis of the contract efficiencies from the client's perspective. The mean, standard deviation, minimum value, maximum value and median of the efficiency scores, relative to the calculations of the standard technical efficiencies, inverted frontier and composite efficiency, are presented.

In order to assess the results of Chart 10.9, an analysis was carried out based on composite technical efficiency (highlighted in bold). It is observed that Contract 9 had the highest mean efficiency (0.873), whereas the lowest mean efficiency was 0.372 in Contract 2. The highest standard deviation in the analysis was 0.218 (Contract 5) and the lowest standard deviation belonged to Contract 7 (0.029). Contract 4 had the DMU with the lowest performance in efficiency (0.149) and Contract 9 had the DMU with the highest performance in efficiency (0.905). The highest median value was that of Contract 9 (0.866), whereas the lowest median value was that of Contract 2 (0.382). The mean efficiency performance of all contracts was 0.658. The standard deviation of all contracts was 0.202 and the median, 0.695.

10.2.2 ANALYSIS OF EFFICIENCY FROM THE PERSPECTIVE OF THE SERVICE PROVIDER

Chart 10.10 presents a consolidated analysis of contract efficiencies from the perspective of the service provider. The mean, standard deviation, minimum value, maximum value and median of the efficiency scores relative to the calculations of the standard technical efficiencies, inverted frontier and composite efficiency are presented.

To assess the results of Chart 10.10, an analysis was performed based on the composite technical efficiency (highlighted in bold). It is observed that Contract 8 had the highest mean performance in efficiency (0.738), with the lowest mean performance in efficiency being 0.206, represented by Contract 4. The highest standard deviation in the analysis was 0.174 (Contract 5) and the lowest standard deviation belonged to Contract 2 (0.027).

Contract 4 had the DMU with the lowest performance in efficiency (0.141) and Contract 5 had the DMU with the highest performance in efficiency (0.873). The highest median value was that of Contract 1 (0.747). The lowest median value was that of Contract 4 (0.204). The mean efficiency performance for all contracts was 0.579, whereas the standard deviation of the contracts was 0.189 and the median, 0.614.

CHART 10.9
Efficiency from the Customer Perspective

Contract	Type of Technical Efficiency	Mean	Standard Deviation	Minimum	Maximum	Median
Contract 1	Standard Efficiency	0.943	0.08	0.766	1	0.979
	Inverted Frontier	0.245	0.015	0.22	0.266	0.242
	Composite Efficiency	**0.849**	**0.044**	**0.752**	**0.89**	**0.864**
Contract 2	Standard Efficiency	0.552	0.076	0.394	0.638	0.562
	Inverted Frontier	0.808	0.111	0.636	1	0.789
	Composite Efficiency	**0.372**	**0.09**	**0.197**	**0.501**	**0.382**
Contract 3	Standard Efficiency	0.951	0.049	0.867	1	0.945
	Inverted Frontier	0.215	0.012	0.2	0.235	0.213
	Composite Efficiency	**0.868**	**0.03**	**0.817**	**0.9**	**0.866**
Contract 4	Standard Efficiency	0.708	0.176	0.298	0.84	0.777
	Inverted Frontier	0.724	0.227	0.345	1	0.753
	Composite Efficiency	**0.492**	**0.175**	**0.149**	**0.748**	**0.521**
Contract 5	Standard Efficiency	0.85	0.25	0.336	1	1
	Inverted Frontier	0.641	0.265	0.309	1	0.596
	Composite Efficiency	**0.605**	**0.218**	**0.202**	**0.845**	**0.693**
Contract 6	Standard Efficiency	0.824	0.04	0.734	0.872	0.83
	Inverted Frontier	0.475	0.117	0.404	0.775	0.432
	Composite Efficiency	**0.675**	**0.057**	**0.533**	**0.734**	**0.688**
Contract 7	Standard Efficiency	0.957	0.039	0.862	1	0.969
	Inverted Frontier	0.979	0.03	0.928	1	1
	Composite Efficiency	**0.489**	**0.029**	**0.431**	**0.532**	**0.489**
Contract 8	Standard Efficiency	0.887	0.057	0.82	1	0.873
	Inverted Frontier	0.481	0.163	0.261	0.648	0.586
	Composite Efficiency	**0.703**	**0.095**	**0.602**	**0.857**	**0.658**
Contract 9	Standard Efficiency	0.997	0.01	0.966	1	1
	Inverted Frontier	0.251	0.082	0.189	0.503	0.228
	Composite Efficiency	**0.873**	**0.041**	**0.748**	**0.905**	**0.886**
General	Standard Efficiency	0.852	0.173	0.298	1	0.895
	Inverted Frontier	0.535	0.292	0.189	1	0.447
	Composite Efficiency	**0.658**	**0.202**	**0.149**	**0.905**	**0.695**

10.2.3   ANALYSIS OF EFFICIENCY IN THE INTEGRATED PERSPECTIVE

Chart 10.11 presents a consolidated analysis of the service contract efficiencies from an integrated perspective. The mean, standard deviation, minimum value, maximum value and median of the efficiency scores relative to the calculations of the standard technical efficiencies, inverted frontier and composite efficiency are presented.

To assess the results of Chart 10.11, an analysis was performed based on composite technical efficiency (highlighted in bold). It was observed that Contract 8 had the highest mean performance in efficiency (0.753), with the lowest mean performance

CHART 10.10
Efficiency from the Perspective of the Service Provider

Contract	Type of Technical Efficiency	Mean	Standard Deviation	Minimum	Maximum	Median
Contract 1	Standard Efficiency	0.852	0.121	0.579	1	0.894
	Inverted Frontier	0.388	0.038	0.304	0.452	0.389
	Composite Efficiency	**0.732**	**0.07**	**0.598**	**0.848**	**0.747**
Contract 2	Standard Efficiency	0.978	0.032	0.916	1	0.996
	Inverted Frontier	0.539	0.029	0.481	0.595	0.537
	Composite Efficiency	**0.719**	**0.027**	**0.66**	**0.76**	**0.726**
Contract 3	Standard Efficiency	0.739	0.103	0.577	0.907	0.719
	Inverted Frontier	0.785	0.151	0.607	1	0.721
	Composite Efficiency	**0.477**	**0.113**	**0.288**	**0.609**	**0.502**
Contract 4	Standard Efficiency	0.367	0.039	0.282	0.418	0.373
	Inverted Frontier	0.956	0.046	0.868	1	0.966
	Composite Efficiency	**0.206**	**0.037**	**0.141**	**0.271**	**0.204**
Contract 5	Standard Efficiency	0.924	0.104	0.703	1	0.959
	Inverted Frontier	0.574	0.27	0.253	0.966	0.562
	Composite Efficiency	**0.675**	**0.174**	**0.42**	**0.873**	**0.689**
Contract 6	Standard Efficiency	0.809	0.094	0.693	1	0.837
	Inverted Frontier	0.807	0.089	0.726	1	0.771
	Composite Efficiency	**0.501**	**0.081**	**0.353**	**0.619**	**0.53**
Contract 7	Standard Efficiency	0.896	0.07	0.785	1	0.879
	Inverted Frontier	0.97	0.04	0.874	1	0.987
	Composite Efficiency	**0.463**	**0.047**	**0.393**	**0.53**	**0.456**
Contract 8	Standard Efficiency	0.937	0.066	0.792	1	0.952
	Inverted Frontier	0.461	0.039	0.417	0.531	0.445
	Composite Efficiency	**0.738**	**0.037**	**0.643**	**0.785**	**0.745**
Contract 9	Standard Efficiency	0.962	0.063	0.841	1	1
	Inverted Frontier	0.56	0.063	0.475	0.682	0.552
	Composite Efficiency	**0.701**	**0.052**	**0.6**	**0.762**	**0.713**
General	Standard Efficiency	0.829	0.196	0.282	1	0.894
	Inverted Frontier	0.671	0.23	0.253	1	0.634
	Composite Efficiency	**0.579**	**0.189**	**0.141**	**0.873**	**0.614**

in efficiency being 0.209, represented by Contract 4. The highest standard deviation in the analysis was 0.147 (Contract 5), with the lowest standard deviation belonging to Contract 8 (0.023).

Contract 4 had the DMU with the lowest performance in efficiency (0.146) and Contract 1 had the DMU with the highest performance in efficiency (0.835). The highest median value was in Contract 8 (0.762), whereas the lowest median value was that of Contract 4 (0.199). The mean performance across all contracts was 0.579, with the corresponding standard deviation of all contracts being 0.178 and the median, 0.617.

CHART 10.11
Efficiency from an Integrated Perspective

Contract	Type of Technical Efficiency	Mean	Standard Deviation	Minimum	Maximum	Median
Contract 1	Standard Efficiency	0.928	0.066	0.759	1	0.931
	Inverted Frontier	0.457	0.057	0.331	0.559	0.452
	Composite Efficiency	**0.735**	**0.056**	**0.6**	**0.835**	**0.738**
Contract 2	Standard Efficiency	0.983	0.03	0.918	1	1
	Inverted Frontier	0.573	0.032	0.527	0.642	0.568
	Composite Efficiency	**0.705**	**0.027**	**0.638**	**0.736**	**0.71**
Contract 3	Standard Efficiency	0.864	0.086	0.725	1	0.88
	Inverted Frontier	0.849	0.129	0.644	1	0.809
	Composite Efficiency	**0.508**	**0.092**	**0.363**	**0.632**	**0.511**
Contract 4	Standard Efficiency	0.385	0.042	0.291	0.439	0.388
	Inverted Frontier	0.967	0.038	0.89	1	0.983
	Composite Efficiency	**0.209**	**0.035**	**0.146**	**0.265**	**0.199**
Contract 5	Standard Efficiency	0.949	0.066	0.804	1	0.977
	Inverted Frontier	0.793	0.259	0.379	1	0.94
	Composite Efficiency	**0.578**	**0.147**	**0.414**	**0.81**	**0.511**
Contract 6	Standard Efficiency	0.882	0.093	0.707	1	0.884
	Inverted Frontier	0.811	0.086	0.763	1	0.771
	Composite Efficiency	**0.535**	**0.076**	**0.373**	**0.618**	**0.555**
Contract 7	Standard Efficiency	0.924	0.071	0.798	1	0.936
	Inverted Frontier	0.973	0.04	0.874	1	0.987
	Composite Efficiency	**0.476**	**0.048**	**0.399**	**0.534**	**0.474**
Contract 8	Standard Efficiency	0.975	0.032	0.913	1	0.989
	Inverted Frontier	0.468	0.038	0.428	0.541	0.453
	Composite Efficiency	**0.753**	**0.023**	**0.71**	**0.783**	**0.762**
Contract 9	Standard Efficiency	0.991	0.027	0.908	1	1
	Inverted Frontier	0.573	0.064	0.486	0.695	0.567
	Composite Efficiency	**0.709**	**0.039**	**0.623**	**0.757**	**0.717**
General	Standard Efficiency	0.876	0.189	0.291	1	0.937
	Inverted Frontier	0.718	0.219	0.331	1	0.723
	Composite Efficiency	**0.579**	**0.178**	**0.146**	**0.835**	**0.617**

10.2.4 COMPARATIVE ANALYSIS OF THE EFFICIENCIES

Chart 10.12 presents a comparative analysis of the client, service provider and integrated efficiencies. The mean, standard deviation, minimum value, maximum value and median composite technical efficiency scores for efficiency calculations from the client's perspective, and those of the service provider and the integrated form, are presented.

Based on the data presented in Chart 10.12, it can be seen that the mean efficiency from the client perspective had a performance of 0.658, higher than the mean efficiency of the service provider and the integrated efficiency (0.579). The

CHART 10.12
Comparative Analysis of the Efficiencies

Contract	From perspective of composite technical efficiency	Median	Standard Deviation	Minimum	Maximum	Median
Contract 1	Client efficiency	0.849	0.044	0.752	0.89	0.864
	Service provider efficiency	0.732	0.07	0.598	0.848	0.747
	Integrated efficiency	**0.709**	**0.039**	**0.623**	**0.757**	**0.717**
Contract 2	Client efficiency	0.372	0.09	0.197	0.501	0.382
	Service provider efficiency	0.719	0.027	0.66	0.76	0.726
	Integrated efficiency	**0.753**	**0.023**	**0.71**	**0.783**	**0.762**
Contract 3	Client efficiency	0.868	0.03	0.817	0.9	0.866
	Service provider efficiency	0.477	0.113	0.288	0.609	0.502
	Integrated efficiency	**0.476**	**0.048**	**0.399**	**0.534**	**0.474**
Contract 4	Client efficiency	0.492	0.175	0.149	0.748	0.521
	Service provider efficiency	0.206	0.037	0.141	0.271	0.204
	Integrated efficiency	**0.535**	**0.076**	**0.373**	**0.618**	**0.555**
Contract 5	Client efficiency	0.605	0.218	0.202	0.845	0.693
	Service provider efficiency	0.675	0.174	0.42	0.873	0.689
	Integrated efficiency	**0.578**	**0.147**	**0.414**	**0.81**	**0.511**
Contract 6	Client efficiency	0.675	0.057	0.533	0.734	0.688
	Service provider efficiency	0.501	0.081	0.353	0.619	0.53
	Integrated efficiency	**0.209**	**0.035**	**0.146**	**0.265**	**0.199**
Contract 7	Client efficiency	0.489	0.029	0.431	0.532	0.489
	Service provider efficiency	0.463	0.047	0.393	0.53	0.456
	Integrated efficiency	**0.508**	**0.092**	**0.363**	**0.632**	**0.511**
Contract 8	Client efficiency	0.703	0.095	0.602	0.857	0.658
	Service provider efficiency	0.738	0.037	0.643	0.785	0.745
	Integrated efficiency	**0.705**	**0.027**	**0.638**	**0.736**	**0.71**
Contract 9	Client efficiency	0.873	0.041	0.748	0.905	0.886
	Service provider efficiency	0.701	0.052	0.6	0.762	0.713
	Integrated efficiency	**0.735**	**0.056**	**0.6**	**0.835**	**0.738**
General	Client efficiency	0.658	0.202	0.149	0.905	0.695
	Service provider efficiency	0.579	0.189	0.141	0.873	0.614
	Integrated efficiency	**0.579**	**0.178**	**0.146**	**0.835**	**0.617**

mean difference between the performance of the client, the service provider and the integrated efficiency was 0.079 (12%). The standard deviation of the efficiencies calculated from the client's perspective was 0.202, compared with the efficiencies calculated for the service provider, 0.189, and with the efficiencies calculated in the integrated model, 0.178.

Regarding the minimum and maximum efficiencies analyzed in Chart 10.12, it is possible to identify that the maximum score obtained in the client efficiency analysis was 0.905, higher than the provider and integrated maximum efficiency scores, that obtained efficiencies of 0.873 and 0.835, respectively. The minimum score obtained in the client efficiency analysis was 0.149, higher than the minimum service provider score of 0.141, and higher than the minimum score of the integrated model, 0.146. It should be noted that, for both the efficiency of the client and the efficiency of the integrated service provider, Contract 4 was the one that presented the lowest efficiency score in the analysis.

10.2.5 ANALYSIS OF VARIANCE OF THE EFFICIENCY MEANS

To perform the statistical tests, ANOVA was used. The data used for the ANOVA test were the longitudinal scores of the composite technical efficiency of service contracts from the perspective of the client, service provider and integrated. For the execution of the ANOVA test, it is necessary that assumptions be met. The assumptions indicated that the data distribution approximated to normality (normality test) and that the data are homogeneous (Levene's homogeneity test). The Kolmogorov-Smirnov test was performed to test the normality of the data. To verify the homogeneity of the data, Levene's test was used. Chart 10.13 presents the validation tests of the ANOVA assumptions.

Chart 10.13 shows that data on the composite efficiencies of the client service contracts are considered to approximate to normality. The normality of the data can be confirmed because the Kolmogorov-Smirnov normality test results presented a probability (P) of 0.586, greater than the threshold required by the test, which is 0.05. Data on the composite efficiencies of service contracts from the provider and integrated perspectives, can also be considered to be normal. Data normality is confirmed by the significance of the test result from the provider and integrated perspectives, which were 0.777 and 0.721, respectively (both higher than the 0.05 threshold required by the test).

In relation to the homogeneity of the variance data, $P = 0.190$ was obtained in the results of Levene's test, performed for the data set of composite technical efficiencies of the client, service provider and the integrated perspectives. In this sense, it can be confirmed that the assumptions for the ANOVA test were fulfilled.

CHART 10.13
Assumptions for Using ANOVA

Data	Kolmogorov-Smirnov (P)	Levene (P)
Service contracts - client	0.586	–
Service contracts - service provider	0.777	–
Service contracts - integrated	0.721	–
Service contracts - total	–	0.19

The ANOVA test was run in three phases. In the first phase, the mean of the client's efficiencies was compared with the mean of the service provider's efficiencies. In the second, the mean of the client's efficiencies was compared with the mean of the efficiencies of the integrated model. In phase three, the mean of the service provider's efficiencies was compared with the mean of the efficiencies of the integrated model Chart 10.14.

The results obtained with the ANOVA test indicated that the mean of the composite technical efficiencies from the client perspective was 0.658, while the mean of the composite technical efficiencies from the perspective of the service provider and the integrated form was 0.579. The resulting difference between the composite technical efficiencies of the client and the service provider was 0.079, while the resulting difference between the composite technical efficiencies of the service provider and the integrated model was zero. In phase one, the F value obtained in the ANOVA test was 8.853, corresponding to a significant P value of 0.003, within the threshold of 0.05 required by the test. The results obtained in the first phase of the ANOVA analysis, for the values of F and P, confirm that the mean efficiency of the service contracts from a client perspective was significantly different from the mean efficiency of service contracts from the perspective of the service provider.

In phase two, the F value obtained in the ANOVA test was 9.400, and the P value was 0.002 (within the 0.05 threshold required by the test). The results obtained in the second phase of the ANOVA analysis confirm that the mean efficiency of service contracts from a client perspective was significantly different from the mean efficiency of service contracts from an integrated perspective.

In the third and last phase of the ANOVA test, the F value obtained was 0.000, and the P value was 0.9914 (outside the 0.05 threshold required by the test). The results obtained in the third phase of the ANOVA analysis confirm that the mean efficiency of service contracts from the service provider perspective was not significantly different from the mean contract service efficiency from the integrated perspective.

Thus, the analysis of the variables associated with the provision of services will be executed in the perspectives of the client and the service provider, since the variables affecting the efficiency of the provider will be the same as those that affect the efficiency of the integrated model.

CHART 10.14
Analysis of Variance (ANOVA)

Phase	Data	Mean of the efficiencies	F	p-value
1	Service contracts – client	0.658	8.853	0.003
	Service contracts – service provider	0.579		
2	Service contracts – client	0.658	9.400	0.002
	Service contracts – integrated	0.579		
3	Service contracts – service provider	0.579	0.000	0.991
	Service contracts – integrated	0.579		

10.2.6   ANALYSIS OF VARIABLES REGARDING EFFICIENCY IN SERVICE PROVISION

After the efficiency analysis, a comparative assessment was made of the influence that the operational characteristics of the clients' fleets have on the efficiency performance of the service contracts. The comparative analysis considers an explanatory model of efficiencies for the client and the service provider. It should be remembered that this assessment will not be performed for the integrated model, since the mean of the efficiencies of this model did not present a significant statistical difference from the mean of the service provider's efficiencies

Initially, tests were made of the assumptions for the Tobit regression. For normality analysis, the Kolmogorov-Smirnov test was used. For the homoscedasticity test of the residues, the Pesarán test was used. The Durbin-Watson test was used to examine the data for the absence of serial autocorrelation and to verify the multicollinearity among the independent variables of the VIF test. The analysis of the tests of assumptions of the Tobit regression showed that the required parameters were exhibited, thus enabling the use of the Tobit regression.

In addition, the efficiency analysis model from the perspective of the client showed an adjusted R^2 of 0.790, and the analysis from the perspective of the services generated an adjusted R^2 of 0.834. Therefore, it is concluded that the models developed have good explanatory power. Chart 10.15 shows the results obtained. It is emphasized that the objective of the comparative analysis was to identify

CHART 10.15
Comparative Analysis between the Models

Operational characteristics	Client		Service provider	
	Influence on efficiency	Sign.	Influence on efficiency	Sign.
Type of Part: Use of parallel part	−17.60%	**0.000**	−12.80%	**0.001**
Type of Network: Use in Dealership	−5.80%	0.328	11.70%	**0.011**
Use of the Fleet: Severe Use	−21.50%	**0.005**	−16.30%	**0.005**
Preventive Maintenance	0.10%	0.776	−0.20%	0.198
Mean Age of the Fleet	−0.10%	**0.014**	0.40%	**0.000**
FIAT	1.30%	0.809	−5.60%	0.173
FORD	−5.70%	0.121	−1.40%	0.617
GM	−9.80%	**0.021**	9.90%	**0.002**
HONDA	11.00%	**0.052**	−3.00%	0.478
MERCEDES	−4.00%	0.409	2.70%	0.454
OUTROS	1.20%	0.896	8.80%	0.223
TOYOTA	−0.20%	**0.003**	8.70%	**0.001**
VOLVO	−10.40%	0.060	−5.40%	0.196
VW	8.80%	0.086	4.80%	0.210
YAMAHA	−15.60%	0.090	22.70%	**0.002**
PEUGEOT	7.10%	0.241	0.10%	0.984
RENAULT	8.70%	**0.012**	−0.80%	0.758

whether the impacts of the operational characteristics were different between the two models.

In analyzing Chart 10.15, it is possible to identify that the use of parallel parts is statistically significant (client: 0.000 and provider: 0.001) and negatively impacts the efficiency of the client and the service provider (client: 17.6% and provider: −12.8%). This is because the quality of the parallel parts is inferior when compared to original and genuine parts. The inferiority of the quality of the parallel parts caused a reduction in their durability, generating a greater number of vehicle breakdowns, and, consequently, an increase in the maintenance costs of the clients and a greater demand for service for the service provider.

The use in dealerships had statistical significance only in the explanatory model of efficiency for the service provider (0.011). It was noted that the maintenance carried out in dealerships generated a positive impact on the efficiency of the service provider of 11.7%. The positive impact generated by the use in dealerships occurred because the maintenance cost in the dealerships was 24.6% higher than the cost of maintenance performed in multi-make workshops. The positive difference between the cost of maintenance in the dealerships and multi-makes generates an increase in revenue for the service provider that is based on the volume of maintenance expenses in the network of commercial establishments.

The variable severity in the use of the fleet was statistically significant and negatively impacted the efficiency of the client (−21.50%) and the provider (−16.30%). It was found that the negative impact generated by the extreme use of the fleet was associated with an increase in maintenance costs. The increase in maintenance costs was generated by the greater number of breakdowns in fleets subjected to extreme use when compared to fleets with normal or moderate use. The increase in costs generates inefficiency for the client, as it negatively impacts the cost per kilometer indicator (US$/km). The increase in fleet losses negatively impacts the service provider, as it generates an increase in operational demands.

In addition to being statistically significant, the mean age of the fleet influenced the efficiencies of the client and the service provider. For the client, the influence of the age of the fleet was negative (−0.10%), since it is concluded that the greater the age of a vehicle, the more costs the vehicle generates with respect to maintenance. For the service provider, the age of the fleet had a positive influence (0.40%), since it was concluded that vehicles of greater age generated higher maintenance costs than new ones. The increase in the expenses on old vehicles increased the revenue of the service provider with commercial establishments.

The GM make exhibited statistical significance, negatively impacting the client's efficiency by −9.80% and positively affecting the provider's efficiency by 9.9%. The negative impact on the client and the positive impact on the service provider is associated with the mean value of parts and services of GM vehicle models. Through an analysis of the data collection bases, it was identified that the mean cost of GM parts and services is 10.2% higher than the cost of parts and services of models belonging to other makes. The increase in costs generates inefficiency for the client, and efficiency for the provider who acquires additional revenue from the network of establishments.

The Honda make exhibited statistical significance only in the explanatory model of efficiency for the client. Honda positively influenced client efficiency by 11%, due to the mean value of its parts and services being 11.2% lower than the mean value of parts and services of its main competitor, Yamaha.

The Toyota make has statistical significance for the explanatory model of efficiency for the client and for the explanatory model of efficiency for the service provider. The impact of Toyota was negative for the efficiency of the client (−10.20%) and positive for the efficiency of the service provider (8.7%). The predominance of Hilux vehicles in the Toyota fleet (92.4%) means that client maintenance costs are higher when compared to maintenance costs for light family vehicles (cars and motorcycles). The increase in costs for clients generates a negative impact on output US$ per km, and, consequently, reduces efficiency. The increase in the cost of the fleet caused by the predominance of the Hilux model generated an increase in revenue for the service provider, remembering that the service provider is remunerated for the amount spent on maintenance in commercial establishments.

In addition to having statistical significance only for the explanatory model of efficiency for the service provider, the Yamaha make generated a beneficial impact on efficiency (22.7%). Yamaha's positive impact on the efficiency of the service provider was associated with the lower price of its parts compared with those of Honda (11.2%). Again, the increase in prices and the stability of demand provided additional revenue for the service provider with a network of commercial establishments.

The Renault make exhibited statistical significance only for the explanatory model of efficiency for the client. Renault generated a beneficial effect on client efficiency of 8.7%, as the costs of its parts and services are reduced due to the negotiation of scale with commercial establishments. Scale negotiations are possible because 86.1% of the Renault models are Sanderos and Logans. The scale negotiations provided a mean discount of 14.9% on the original value of parts and services, and, consequently, helped to reduce the cost per kilometer of clients' fleets.

The variable, preventive maintenance, for Fiat, Ford, Mercedes, Volvo, Volkswagen, Peugeot and other makes did not exhibit statistical significance in the regression analysis of the explanatory models of the efficiencies for the client and service provider. The variables used in Yamaha dealerships exhibited no statistical significance, with the exception of the explanatory model of the efficiency for the client. The variables Honda and Renault exhibited statistical significance only in the explanatory model of efficiency for the service provider.

10.3 EFFICIENCY AND BENCHMARKING: CASE STUDY IN A FUEL STATION NETWORK

In this section, we present the results regarding the practical problems of productivity and efficiency management discussed in Section 3.3. First, we will briefly describe the working method adopted in the research, and, later, the results achieved.

10.3.1 WORKING METHOD

This research was carried out in operational units of a network composed of five fuel retailers (called Stations 1, 2, 3, 4 and 5). Therefore, the unit of analysis for this work is the network of fuel stations. The stations offer five types of fuel: common gasoline,

gasoline with additive(s), S500 diesel, S10 diesel and ethanol. These five products represent 95% of the organization's revenue, which enabled a fuel-focused analysis. The standard technical efficiency was used for analysis. To assist in the development of the work, the fuel station network made available a team of professionals.

A longitudinal survey was carried out considering the years 2014, 2015 and 2016 (partial), and a transverse survey was carried out, comparing the five stations. Thus, the data analyzed are presented in a data panel. The monthly volume of fuel sold (liters) at each of the five stations was determined as a DMU. Thus, the total obtained was 145 DMUs. The use of the CRS model with input orientation was defined. Initially, based on the literature, a pre-listing of variables for the DEA model was carried out, to be discussed and validated later together with the specialists who supported the development of the study. Chart 10.16 presents the final listing of the variables used in the model, as well as their functions (input or output) and the unit of measurement used in the collection.

ANOVA and its assumptions (Shapiro-Wilk and Bartlett) were also used to statistically test the differences in the means of the efficiency scores of the fuel stations analyzed.

10.3.2 RESULTS

Chart 10.17 shows the efficiencies of the five fuel retailers over 29 months, in chronological order. In addition, the mean monthly efficiencies and each unit of analysis are presented. The monthly minimum and maximum efficiencies and that for each DMU in the period analyzed are also presented. Finally, the mean and standard deviation calculations are presented.

CHART 10.16
Description and Detail of the Inputs and Outputs Used in the DEA Model

Variable	Description	Name	Unit
Input 1	Total volume (l) of fuel bought, contemplating five types of commercialized fuels: common gasoline, gasoline with additives, S500 diesel, S10 diesel and ethanol.	Volume of fuel bought	l
Input 2	Storage capacity (l) of tanks installed underground	Capacity of tanks	l
Input 3	Covered area (m²) to attend clients, where pump attendants and vehicles circulate.	Forecourt area	m²
Input 4	Quantity of pump nozzles (un.) available	Quantity of nozzles	un.
Input 5	Total time (h) station was in operation in the period.	Hours of operation	h
Input 6	Time worked (h) in the period, considering the number of employees and operating hours.	Hours worked by forecourt attendants	h
Input 7	Quantity (un.) of acceptable payment means.	Quantity of payment means	un.
Output1	Total volume (l) of fuel sold, contemplating the five types of fuel sold: common gasoline, gasoline with additives, S500 diesel, S10 diesel and ethanol.	Volume of fuel sold	l

---

CHART 10.17

Efficiency of the Units of Analysis

Month/Year	Efficiency of Station 1	Efficiency of Station 2	Efficiency of Station 3	Efficiency of Station 4	Efficiency of Station 5	Efficiency Mean month	Efficiency Minimum month	Efficiency Maximum month	Media month	Standard Deviation month
Jan/14	0.860	0.862	0.887	0.871	0.887	0.873	0.860	0.887	0.871	0.013
Feb/14	1.000	0.963	0.943	0.891	0.863	0.932	0.863	1.000	0.943	0.055
Mar/14	0.892	0.932	0.936	0.866	0.920	0.909	0.866	0.936	0.920	0.030
Apr/14	0.878	0.928	0.897	0.962	0.820	0.897	0.820	0.962	0.897	0.054
May/14	0.897	0.877	0.923	0.789	0.856	0.868	0.789	0.923	0.877	0.051
Jun/14	0.880	0.884	0.893	0.898	0.852	0.881	0.852	0.898	0.884	0.018
Jul/14	0.870	0.933	0.909	0.830	0.923	0.893	0.830	0.933	0.909	0.043
Aug/14	0.904	0.957	0.913	0.972	0.913	0.932	0.904	0.972	0.913	0.030
Sep/14	0.879	0.886	0.965	0.853	0.810	0.878	0.810	0.965	0.879	0.057
Oct/14	0.889	0.909	0.881	0.828	0.841	0.870	0.828	0.909	0.881	0.034
Nov/14	0.915	0.922	0.963	0.924	0.899	0.925	0.899	0.963	0.922	0.024
Dec/14	0.894	0.906	0.856	0.903	0.904	0.893	0.856	0.906	0.903	0.021
Jan/15	0.859	0.874	0.936	0.785	0.834	0.858	0.785	0.936	0.859	0.055
Feb/15	0.885	0.915	0.885	0.855	0.895	0.887	0.855	0.915	0.885	0.022
Mar/15	0.885	0.937	0.927	0.919	0.824	0.898	0.824	0.937	0.919	0.046
Apr/15	0.936	0.928	0.917	0.929	0.903	0.923	0.903	0.936	0.928	0.013
May/15	0.911	0.926	0.953	0.835	0.891	0.903	0.835	0.953	0.911	0.044
Jun/15	0.901	0.885	0.909	0.837	0.847	0.876	0.837	0.909	0.885	0.032
Jul/15	0.951	0.981	0.949	0.944	0.898	0.945	0.898	0.981	0.949	0.030
Aug/15	0.971	0.971	0.953	0.838	0.872	0.921	0.838	0.971	0.953	0.062
Sep/15	0.929	0.927	0.925	0.820	0.847	0.890	0.820	0.929	0.925	0.052

(Continued)

CHART 10.17 (CONTINUED)
Efficiency of the Units of Analysis

Month/Year	Efficiency of Station 1	Efficiency of Station 2	Efficiency of Station 3	Efficiency of Station 4	Efficiency of Station 5	Efficiency Mean month	Efficiency Minimum month	Efficiency Maximum month	Media month	Standard Deviation month
Oct/15	1.000	1.000	1.000	0.908	0.888	0.959	0.888	1.000	1.000	0.056
Nov/15	1.000	0.979	0.986	0.791	0.893	0.930	0.791	1.000	0.979	0.088
Dec/15	0.993	1.000	0.991	0.926	0.828	0.948	0.828	1.000	0.991	0.073
Jan/16	0.964	0.949	1.000	0.918	0.917	0.949	0.917	1.000	0.949	0.035
Feb/16	0.959	0.910	0.946	0.977	0.923	0.943	0.910	0.977	0.946	0.027
Mar/16	0.904	0.910	0.929	0.846	0.849	0.888	0.846	0.929	0.904	0.038
Apr/16	0.936	0.939	0.944	0.906	0.851	0.915	0.851	0.944	0.936	0.039
May/16	0.928	0.937	0.942	0.832	0.850	0.898	0.832	0.942	0.928	0.052
Mean Station efficiency	0.920	0.929	0.933	0.878	0.872	0.906	0.849	0.949	0.919	0.041
Minimum station efficiency	0.859	0.862	0.856	0.785	0.810	0.858	0.785	0.887	0.859	0.013
Maximum station efficiency	1.000	1.000	1.000	0.977	0.923	0.959	0.917	1.000	1.000	0.088
Station mean	0.904	0.928	0.936	0.871	0.872	0.898	0.846	0.942	0.919	0.039
Station standard deviation	0.044	0.037	0.036	0.055	0.034	0.028	0.037	0.034	0.036	0.018

When analyzing Chart 10.17, it is understood that the greater the performance of a given station in a certain month, the higher the efficiency score resulting from the calculation performed in DEA. Thus, it is observed that the highest efficiency performances referred to Station 1 (Feb/14), Station 1 (Oct/15), Station 1 (Nov/15), Station 2 (Oct/15), Station 2 (Dec/15), Station 3 (Oct/15) and Station 3 (Jan/16).

The DMUs referring to these stations, in these months, obtained an efficiency equal to 1.00. It can be seen that the stations with the highest performances are Stations 1, 2 and 3, and that there is a period, between October 2015 and January 2016, in which these units had their highest performances. With regard to the lowest efficiency performance, this occurred in Station 4 in January 2015. Subsequently, the poorest performances were Station 4 in May/14 and Station 4 in Nov/15. The 25 lowest efficiency performances referred to Stations 4 and 5. It can be observed that Station 1 reached maximum efficiency in three months, but, among the low efficiencies, Station 2 was considered to have the lowest efficiency, achieving an efficiency of 0.862 in its worst month of the analysis period.

The performance of the three most efficient stations in relation to the mean of the five stations during the 29 months can be observed. Station 1 was above the mean for 69.9% of the period (20 months), while Station 3 was above the mean for 83.7% of the period (24 months). The station that remained above average for the longest period was Station 2, for 86.2% of the period (25 months). It can be noticed that Stations 1 and 3 had their months of instability in relation to the network throughout the entire year of 2014 until May 2015, whereas Station 2 did not exhibit the same behavior and still had two months of poor performance in relation to the others in 2016 (January and February).

In relation to the two stations with lower efficiency scores, the same analyses were carried out as performed with the group of the more efficient stations. Thus, for maximum and minimum efficiencies, Station 4 exhibited 1a) maximum efficiency: 0.977 (Feb/16) and 1b) minimum efficiency: 0.785 (Jan/15); whereas Station 5 exhibited 2a) maximum efficiency: 0.923 (Feb/16) and 2b) minimum efficiency: 0.810 (Sep/14). Both exhibited maximum efficiency in February 2016. It was also determined that the difference between the minimum and the maximum efficiency in the period presented by Station 4 was greater than that presented by Station 5. Thus, Station 4 presented the highest maximum efficiency score and the lowest minimum efficiency score in the group of the less-efficient stations.

The performances of Stations 4 and 5 can be compared with the mean of the five stations. Station 4 is above the mean for only 24.1% of the period (7 months) and Station 5 is above the mean for 17.2% of the period (5 months). It can be seen that the predominant period in which Stations 4 and 5 performed below the mean occurred from May 2015, exactly the opposite of the behavior of Stations 1 and 3. In the months prior to May 2015, there were fluctuations between months that were more or less efficient than the mean of the network.

A comparison of the station efficiencies was also made in relation to its own mean in the analysis period. This comparison was carried out for the stations with higher efficiency scores. Station 3 was the only one with more than 50% of the number of months with efficiency above its own mean value. Station 3 was above its mean efficiency for 51.7% of the months (15 months).

Station 1 exhibited above-average efficiency for 12 months, which represents 41.4% of the period analyzed, whereas Station 2 spent 13 months with above-average efficiency, representing 44.8% of the period under analysis. The months with efficiencies below the mean were concentrated between January 2014 and June 2015; that is, after the second half of 2015, an increase in performance in this station was observed.

Comparisons were then made of the efficiencies of the less efficient stations with respect to their mean in the period of analysis. Stations 4 and 5 had the same result, with 48.3% of the period analyzed being above their mean efficiency, representing 14 months. It can be observed that, in these stations, there was no period of concentration of the months below the mean efficiency, since the low-efficiency months occurred until the last month of the period of analysis. Thus, it is understood that these stations did not achieve a significant increase in performance over the period of analysis.

10.3.3　Statistical Analyses

To perform analysis of the variance on the efficiency scores, they were divided into groups of 29 DMUs corresponding to the 29 months of analysis of each station. Thus, there are five groups:

(i) DMUs from 1 to 29, corresponding to Station 1.
(ii) DMUs from 30 to 58, corresponding to Station 2.
(iii) DMUs 59 to 87, corresponding to Station 3.
(iv) DMUs 88 to 116, corresponding to Station 4.
(v) DMUs 117 to 145, corresponding to Station 5.

As for the ANOVA statistical tests and their assumptions, it was verified that the result of the Kolmogorov-Smirnov test for Station 1 ($P = 0.827$), Station 2 ($P = 0.584$), Station 3 ($P = 0.466$), Station 4 ($P = 0.604$) and Station 5 ($P = 0.773$) presented a level of significance higher than 0.05. Therefore, it is confirmed that the data from all five stations approximated to a normal distribution.

Furthermore, it was found that the result obtained for the Bartlett test ($P = 0.846$) was greater than the 0.05 threshold. Therefore, it is confirmed that the variances of the data were homogeneous. Thus, the analysis of assumptions allowed use of the ANOVA test. The results for the ANOVA test are displayed in Chart 10.18.

In the results of Chart 10.18, it is observed that the value of F is equal to 13.898235 and the P value is equal to 0.0000000014. The higher the F score, the more significant is the P value of the ANOVA test. Thus, it can be affirmed that the observed difference between the means of the efficiencies of the stations was statistically significant. Therefore, it can be affirmed that it is significant to analyze the difference between the technical efficiencies of the five retail points. Finally, an analysis of targets and slack was made to identify the improvement targets for each fuel station analyzed.

CHART 10.18

ANOVA Test Among the Stations Analyzed

Source of variation	SQ	gl	MQ	F	p-value	Critical F
Among groups	0.09754046	4	0.0243851	13.89823528	0.0000000014	2.436317464
Within groups	0.24563666	140	0.0017545	–	–	–
Total	**0.34317712**	**144**	0.0261396			

10.4 EFFICIENCY AND INVESTMENT PROJECTS: CASE STUDY IN A PETROCHEMICAL COMPANY

In this section, we present the results regarding the practical problem of productivity and efficiency management discussed in Section 3.4. First, we will briefly describe the working method adopted in the research and, later, the results achieved.

10.4.1 WORKING METHOD

The research was conducted on a petrochemical company. The plant, or production unit, is responsible for the transformation of resources into finished products. In addition to the finished products, the process generates waste, which may be solid, liquid or gaseous. Some petrochemical processes produce materials with distinct physico-chemical characteristics. These are called grades, which are produced in batch production campaigns. These campaigns of grade production were considered to be the DMUs.

The DMUs were grouped and analyzed in two distinct situations. The first grouping was formed, taking into account the period between the start of a project (project A), and the end, which represents the start of the next project (project B). This period will be called period 1 and so on. The projects are characterized by the technical changes they cause in the production process, and the purpose of the grouping is to determine whether, within the period analyzed, efficiency changes occur, these being related, or not, to the process of learning and continuous improvement.

A longitudinal survey was conducted covering the years 2004 to 2011, segregated into periods according to Chart 10.19. The CRS and VRS models were used together, along with input orientation.

Initially, with the support from the literature, a pre-listing of variables for the DEA model was performed, which was later discussed and validated together with the specialists supporting the study development. Chart 10.20 shows the final listing of the variables used in the model, as well as their function (input or output) and the unit of measurement used in the collection.

ANOVA, the t-test, the Kruskal-Wallis test and the Malmquist Index were also used to complement the analyses performed.

CHART 10.19

Periods of Analysis

Period of analysis	Start	End	Duration (years)
Period 1	07/02/2004	16/07/2006	2.4
Period 2	25/07/2006	01/08/2009	3.0
Period 3	01/08/2009	31/10/2011	2.3

CHART 10.20

Details of Inputs and Outputs Used

Variable	Product	Unit
Input 1	Time	h
Input 2	Energy + Steam	kJ
Input 3	Ethanol	Nm^3
Input 4	Propane	Nm^3
Input 5	Hexane	m^3
Output	Rubber	T

10.4.2   RESULTS

The results were assessed by analysis period. Each analysis period uses the results of standard, inverted, composite and scale technical efficiency calculations. For the analysis of the efficiencies of the periods, the VRS model was used.

10.4.3   ANALYSIS OF PERIOD 1 (FEB/2004 TO JUL/2006)

Chart 10.21 presents the results of the calculation of the standard, inverted, composite and scale efficiencies for the 25 G1 grade campaigns and 18 G2 grade campaigns in the first analysis period. The calculated values were used to analyze the performance in relation to the efficiency of each analysis unit and the period as a whole.

The DMUs are listed in chronological order. For the G1 grade, it is observed that the DMUs G1P1C02, G1P1C13 and G1P1C17 present the highest efficiencies; the three efficient DMUs represent 12% of the total units of analysis. The DMUs G1P1C07, G1P1C08, G1P1C09 and G1P1C20 are the DMUs that present the lowest efficiencies observed in the sample. Based on these observations, it can be inferred that efficiency is not an indicator under control, and no effects of the learning process and continuous improvement are observed.

For the G2 grade, the DMUs G2P1C09 and G2P1C14 have the highest efficiency relation ratios. The DMUs G2P1C01, G2P1C03, G2P1C04, G2P1C10 and G2P1C13 are the DMUs that have the lowest observed efficiencies in the sample. It is also noted that the DMUs that have the lowest efficiencies are located more toward the beginning of the period. Based on this observation, it is possible to infer that, in the

CHART 10.21

Efficiencies (VRS) Standard, Inverted, Composite and Scale Efficiencies in Grades G1 and G2 in Period 1

		Grade G1 Period 1					Grade G2 Period 1		
DMUs	VRS	Inverted	Composite	Efficiency of scale	DMUs	VRS	Inverted	Composite	Efficiency of scale
G1P1C01	0.9221	0.9097	0.5062	0.9706	G2P1C01	0.8409	1.0000	0.4205	0.9937
G1P1C02	1.0000	0.8441	0.5779	1.0000	G2P1C02	1.0000	0.9256	0.5372	0.9861
G1P1C03	0.9314	1.0000	0.4657	0.9516	G2P1C03	0.9373	1.0000	0.4686	0.8443
G1P1C04	0.8677	0.9555	0.4561	0.9764	G2P1C04	0.8743	1.0000	0.4371	0.9136
G1P1C05	0.9416	0.9274	0.5071	0.9679	G2P1C05	1.0000	1.0000	0.5000	0.9783
G1P1C06	0.9194	0.9573	0.481	0.9616	G2P1C06	0.9347	0.9701	0.4823	0.9331
G1P1C07	0.9194	1.0000	0.4597	0.7429	G2P1C07	0.9418	0.9537	0.4940	0.9814
G1P1C08	0.8049	1.0000	0.4025	0.9843	G2P1C08	0.9437	0.9159	0.5139	0.9733
G1P1C09	0.7992	1.0000	0.3996	0.9810	G2P1C09	1.0000	0.7780	0.6110	1.0000
G1P1C10	0.9185	0.8679	0.5253	0.9818	G2P1C10	0.9274	1.0000	0.4637	0.8437
G1P1C11	0.9670	0.9104	0.5283	0.9242	G2P1C11	0.9693	0.9368	0.5163	0.8821
G1P1C12	0.9578	0.9272	0.5153	0.9574	G2P1C12	0.9357	0.9501	0.4928	0.8963
G1P1C13	1.0000	0.7380	0.631	1.0000	G2P1C13	0.9356	1.0000	0.4678	0.9172
G1P1C14	1.0000	0.9662	0.5169	0.9434	G2P1C14	1.0000	0.9323	0.5339	1.0000
G1P1C15	0.9913	0.8962	0.5475	0.9550	G2P1C15	0.9834	0.9318	0.5258	0.8605
G1P1C16	0.9486	0.8815	0.5336	0.9683	G2P1C16	0.9462	0.9393	0.5035	0.9922
G1P1C17	1.0000	0.8682	0.5659	1.0000	G2P1C17	0.9828	0.9435	0.5197	0.9990
G1P1C18	1.0000	0.9568	0.5216	0.9382	G2P1C18	1.0000	1.0000	0.5000	0.9689
G1P1C19	0.9892	0.8841	0.5526	0.9718	–	–	–	–	–
G1P1C20	0.9043	1.0000	0.4521	0.9682	–	–	–	–	–
G1P1C21	1.0000	1.0000	0.5	0.9493	–	–	–	–	–
G1P1C22	0.9644	0.9058	0.5293	0.9603	–	–	–	–	–
G1P1C23	0.9645	1.0000	0.4823	0.9457	–	–	–	–	–
G1P1C24	0.9716	0.9108	0.5304	0.9501	–	–	–	–	–
G1P1C25	0.9707	1.0000	0.4853	0.9414	–	–	–	–	–

first period of grade G2, an efficiency increase occurred over time, and that this increase may be related to the process of learning and continuous improvement, since, in the period, no technological changes in the process were realized.

10.4.4 ANALYSIS OF PERIOD 2 (JUL/2006 TO AUG/2009)

The second review period began in July 2006 after the implementation of a project to reduce input4 (propane) consumption and ended in August 2009. Chart 10.22 presents the results of the calculations of the standard, inverted, composite and scale efficiencies for the 34 G1 grade campaigns and the 26 G2 grade campaigns in the second analysis period. For the G1 grade, six DMUs, G1P2C30, G1P2C34, G1P2C37, G1P2C39, G1P2C55, and G1P2C58 are shown to have the highest performances.

CHART 10.22

(VRS) Standard, Inverted, Composite and Scale Efficiency Techniques in Grades G1 and G2 in Period 2

	Grade G1 Period 2				Grade G2 Period 2			
DMUs	VRS	Inverted	Composite	Efficiency of scale	DMUs	VRS	Inverted	Composite	Efficiency of scale
G1P2C26	1.0000	0.8802	0.5599	0.9922	G2P2C19	0.9895	0.8183	0.5856	0.9807
G1P2C27	0.9183	0.9303	0.4940	0.9999	G2P2C20	1.0000	0.8829	0.5585	1.0000
G1P2C28	1.0000	0.9550	0.5225	0.9755	G2P2C21	0.9758	0.9409	0.5174	0.9962
G1P2C29	0.9601	0.8982	0.5310	0.9991	G2P2C22	0.9703	0.9226	0.5239	0.9994
G1P2C30	1.0000	0.8790	0.5605	1.0000	G2P2C23	1.0000	0.9219	0.5391	0.9970
G1P2C31	1.0000	0.8859	0.5571	0.9980	G2P2C24	1.0000	0.9739	0.5131	0.9791
G1P2C32	1.0000	0.9300	0.5350	0.9871	G2P2C25	0.9360	0.9900	0.4730	0.9999
G1P2C33	0.9856	0.9078	0.5389	0.9818	G2P2C26	1.0000	1.0000	0.5000	0.9668
G1P2C34	1.0000	0.8479	0.5760	1.0000	G2P2C27	1.0000	0.9352	0.5324	0.9927
G1P2C35	1.0000	0.9817	0.5091	0.8891	G2P2C28	0.9765	0.9370	0.5197	0.9999
G1P2C36	1.0000	1.0000	0.5000	0.7521	G2P2C29	1.0000	1.0000	0.5000	1.0000
G1P2C37	1.0000	0.8916	0.5542	1.0000	G2P2C30	0.9431	0.9718	0.4857	0.9954
G1P2C38	0.9504	0.9321	0.5092	0.9990	G2P2C31	0.9685	0.9439	0.5123	0.9999
G1P2C39	1.0000	0.9283	0.5359	1.0000	G2P2C32	1.0000	0.9947	0.5026	0.9646
G1P2C40	0.9214	1.0000	0.4607	0.9659	G2P2C33	0.9656	1.0000	0.4828	0.9558
G1P2C41	0.8676	1.0000	0.4338	0.9379	G2P2C34	1.0000	1.0000	0.5000	0.7359
G1P2C42	0.9004	1.0000	0.4502	0.9569	G2P2C35	1.0000	1.0000	0.5000	0.9372
G1P2C43	0.9214	0.9471	0.4871	0.9976	G2P2C36	0.9935	0.9606	0.5164	0.9375
G1P2C44	0.9336	0.9381	0.4978	0.9984	G2P2C37	1.0000	0.8692	0.5654	1.0000
G1P2C45	0.9782	0.9161	0.5311	0.9875	G2P2C38	1.0000	0.9202	0.5399	1.0000
G1P2C46	0.9830	0.9167	0.5331	0.9777	G2P2C39	0.9808	0.9513	0.5148	0.9993
G1P2C47	0.9038	1.0000	0.4519	0.9944	G2P2C40	0.9765	1.0000	0.4883	0.9999
G1P2C48	0.9660	0.9239	0.5210	0.9942	G2P2C41	1.0000	0.8077	0.5962	1.0000
G1P2C49	0.9730	0.9295	0.5217	0.9902	G2P2C42	0.9662	0.8910	0.5376	0.9879
G1P2C50	0.9656	0.8787	0.5434	0.9834	G2P2C43	0.9924	0.8329	0.5798	0.9870
G1P2C51	1.0000	0.9224	0.5388	0.9840	G2P2C44	0.8453	1.0000	0.4226	0.9991
G1P2C52	0.9538	0.9257	0.5140	0.9957	–	–	–	–	–
G1P2C53	0.9819	0.9116	0.5351	0.9956	–	–	–	–	–
G1P2C54	0.9401	0.9091	0.5155	0.9977	–	–	–	–	–
G1P2C55	1.0000	0.9135	0.5432	1.0000	–	–	–	–	–
G1P2C56	1.0000	0.8851	0.5574	0.9557	–	–	–	–	–
G1P2C57	0.9702	0.9135	0.5283	0.9844	–	–	–	–	–
G1P2C58	1.0000	0.9047	0.5476	1.0000	–	–	–	–	–
G1P2C59	1.0000	1.0000	0.5000	1.0000	–	–	–	–	–

Comparing the percentage of DMUs exhibiting the higher performance in the first period, it was observed that there was no significant variation. The seven DMUs of the second period represent 17% of the total DMUs, while, in the first period, the best-performing DMUs represent 12% of the sample (3 of 25).

The analysis of the data in Chart 10.22 indicates that the period presents satisfactory results until the fourteenth campaign, with high efficiencies, and four of the six

DMUs having greater efficiencies in this period. After campaign 14 (G1P2C39), the DMUs exhibited low efficiency indexes. At the end of the analysis period, the campaigns once again showed greater efficiencies. The behavior of the units of analysis, observed throughout the period, cannot be explained by learning and continuous improvement. It is concluded that, for the second period of grade G1, as observed in the first period, the process of learning and continuous improvement did not influence efficiency in the sense of raising it over time.

For the G2 grade in the second analysis period, five DMUs G2P2C20, G2P2C29, G2P2C37, G2P2C38 and G2P2C41, exhibited the highest performances. This represents 19% of the analysis units of the period (5 of 26), compared with the result of period 1, in which 11% of the DMUs were efficient, causing a significant increase in the percentage of efficient DMUs. It is also observed that the G2P2C19 and G2P2C43 DMUs, even though they are not considered to be among the most efficient ones, present high performances when considering the composite efficiency that takes into account the inverted frontier; that is, these DMUs present less variability in the partial efficiencies of the inputs.

The least efficient units are G2P2C25, G2P2C30, G2P2C33 and G2P2C44. With regard to the process of learning and continuous improvement, based on the chronology of the units, it cannot be affirmed that it positively influenced this period.

10.4.5 Analysis of Period 3 (Aug/2009 to Nov/2011)

The third period of analysis begins in August 2009, after the implementation of project 2, with the objective of increasing the efficiency of input 1 (time), and ends in November 2011. Chart 10.23 presents, in chronological order, the result of the calculations of the standard, inverted, composite and scaling efficiencies for the 21 G1 grade campaigns and 25 G2 grade campaigns in the third analysis period.

For the G1 grade, six DMUs G1P3C60, G1P3C62, G1P3C64, G1P3C67, G1P3C79 and G1P3C80 are shown to have the highest performances. Comparing the percentage of DMUs with high performance in the first and second periods (12%) with that in the third, an increase was observed with 28.6% of the DMUs (6 of 21) being efficient. Two DMUs not considered efficient, but presenting good performance due to the greater balance between the individual efficiencies of each input, are the DMUs G1P3C63 and G1P3C72.

The DMUs that showed the lowest efficiencies are G1P3C61, G1P3C65, G1P3C66, G1P3C69 and G1P3C74. These DMUs are more concentrated at the beginning of the analysis period, suggesting that, for the G1 grade in the third period, there is a positive influence of the learning process and continuous improvement. The beginning of the period that presents a greater oscillation between efficient and non-efficient DMUs is characteristic of new technologies or changes that present adaptation phases. Inferences with respect to the effectiveness of project 2 cannot be made based on the analyses performed in this section, since there is no basis for comparison between the periods, since the efficiency is calculated independently in each period.

For the G2 grade in the third analysis period (Chart 10.23), two DMUs, G2P355 and G2P367, show the highest performances. It is also noted in Chart 10.23 that the DMUs with the lowest efficiencies are more concentrated at the beginning of

CHART 10.23
Standard, Inverted, Composite and Scale Technical Efficiency (VRS) in Grades G1 and G2 in Period 3

	Grade G1 Period 3					Grade G2 Period 3			
DMUs	VRS	Inverted	Composite	Efficiency of scale	DMUs	VRS	Inverted	Composite	Efficiency of scale
G1P3C60	1.0000	0.9515	0.5243	1.0000	G2P3C45	0.9590	1.0000	0.4795	0.9782
G1P3C61	0.9494	0.9950	0.4772	0.9971	G2P3C46	1.0000	1.0000	0.5000	0.8936
G1P3C62	1.0000	1.0000	0.5000	1.0000	G2P3C47	1.0000	0.9061	0.5470	0.9896
G1P3C63	1.0000	0.7940	0.6030	0.9001	G2P3C48	0.9554	0.9736	0.4909	0.9620
G1P3C64	1.0000	0.9466	0.5267	1.0000	G2P3C49	0.9596	0.9654	0.4971	0.9602
G1P3C65	0.6437	1.0000	0.3218	0.9399	G2P3C50	0.9617	1.0000	0.4809	0.9576
G1P3C66	0.9240	1.0000	0.4620	0.9979	G2P3C51	0.9555	1.0000	0.4777	0.9589
G1P3C67	1.0000	1.0000	0.5000	1.0000	G2P3C52	0.9943	0.9577	0.5183	0.9524
G1P3C68	1.0000	1.0000	0.5000	0.9844	G2P3C53	0.9942	0.9804	0.5069	0.9488
G1P3C69	0.9143	0.9200	0.4971	0.9999	G2P3C54	0.9194	0.9578	0.4808	0.9919
G1P3C70	1.0000	0.9242	0.5379	0.9612	G2P3C55	1.0000	0.8453	0.5774	1.0000
G1P3C71	1.0000	1.0000	0.5000	0.9851	G2P3C56	0.9988	1.0000	0.4994	0.9648
G1P3C72	0.9964	0.7799	0.6083	0.8855	G2P3C57	0.9661	1.0000	0.4830	0.9667
G1P3C73	1.0000	0.9726	0.5137	0.9971	G2P3C58	0.8457	1.0000	0.4229	0.9990
G1P3C74	0.9389	0.9406	0.4991	0.9985	G2P3C59	0.9659	0.9708	0.4976	0.9554
G1P3C75	0.9909	0.9713	0.5098	0.9948	G2P3C60	0.9728	0.9391	0.5169	0.9651
G1P3C76	0.9847	0.9520	0.5164	0.9869	G2P3C61	1.0000	1.0000	0.5000	0.9329
G1P3C77	1.0000	1.0000	0.5000	0.8938	G2P3C62	1.0000	1.0000	0.5000	0.9284
G1P3C78	1.0000	0.7936	0.6032	0.9384	G2P3C63	0.9665	0.9637	0.5014	0.9584
G1P3C79	1.0000	0.6701	0.6649	1.0000	G2P3C64	1.0000	0.9323	0.5338	0.9508
G1P3C80	1.0000	0.8945	0.5528	1.0000	G2P3C65	1.0000	0.9882	0.5059	0.9439
–	–	–	–	–	G2P3C66	0.9768	0.9531	0.5118	0.9751
–	–	–	–	–	G2P3C67	1.0000	0.9230	0.5385	1.0000
–	–	–	–	–	G2P3C68	0.9620	0.9219	0.5200	0.9921
–	–	–	–	–	G2P3C69	0.9755	0.9478	0.5139	0.9608

the analysis period, namely the DMUs G2P3C45, G2P3C48, G2P3C50, G2P3C51, G2P3C54, G2P3C57 and G2P3C58. The concentration of less efficient DMUs at the beginning of the period suggests increases in efficiency over the period, which may be associated with the learning process and continuous improvement.

In summary, the analyses of grades G1 and G2 indicate that the learning process and continuous improvement had a positive influence on the composite efficiency only in periods 1 and 3. Chart 10.24 presents a summary of the conclusions regarding the presence of the learning process and the continuous improvement of the composite efficiency over the three periods of analysis.

To better understand the results observed, the analysis is presented of the periods in terms of the Malmquist Index, which proposes to assess the efficiency variations and dissect them in terms of changes in technical efficiency and technological change.

CHART 10.24
Learning Process in the Periods

Period	Grade G1	Grade G2
1	Observed	Observed
2	Not Observed	Not Observed
3	Observed	Observed

10.4.6 COMPARATIVE ANALYSIS OF THE UNITS OF ANALYSIS – MALMQUIST INDEX (GRADES G1 AND G2)

The periods of analysis are analyzed longitudinally to determine the impact of the projects by comparing the efficiencies among the periods of analysis. The comparison was made using the Malmquist Index and its components: (i) variation of technical efficiency; and (ii) variation of technological change. In the variation of the technical efficiency, it used the technical efficiency of the CRS and VRS models. In addition, the efficiency of scale was also calculated. In the analysis of the variation of the efficiencies, values above 1 represent efficiency gains, while values below 1 represent efficiency losses.

Chart 10.25 presents the summary concerning the analysis of grade G1. It should be noted that, from period 1 to period 2, the CRS technical efficiency gain was 4.1% (of which 3.1% was due to efficiency gains and 1.0% to scale). However, there was a drop in efficiency attributed to technological change, of the order of 6.2%. Thus, the relation of these variations, resulted in a reduction in the total factor productivity (Malmquist Index) of 2.1%.

The assessment of the mean variation from period 2 to period 3 indicates that there was a decrease in the CRS technical efficiency of 2.8% (i.e., 1 − 0.972), but there was a technological evolution in the sector of the order of 7.2%. This resulted in an increase in total factor productivity (Malquist Index) of 4.2%.

By means of longitudinal analysis of grade G1, it is concluded that project 1 was not positive for the company or process in terms of efficiency. On the other hand, project 2 was positive for having raised the efficiency frontier, and also for identifying the opportunity to increase technical efficiency through the process of learning and continuous improvement.

Chart 10.26 presents the summary of grade G2. Between the first and second periods (1–2), the CRS technical efficiency gain was 8.25% (of which 3.63% was

---

CHART 10.25
Summary of Efficiency Variation - Index Malmquist Grade G1

Period	Variation in technical efficiency CRS	Technological Change	Variation in technical efficiency VRS	Efficiency of Scale (CRS/VRS)	Malmquist Index
1–2	1.0410	0.9380	1.0310	1.0100	0.9790
2–3	0.9720	1.0720	0.9710	1.0010	1.0420

CHART 10.26

Summary Variation Efficiency Index Malmquist Grade G2

Period	Variation in technical efficiency CRS	Technological Change	Variation in technical efficiency VRS	Efficiency of Scale (CRS/VRS)	Malmquist Index
1–2	1.0825	0.9198	1.0462	1.0363	1.0042
2–3	0.9840	1.0545	0.9923	0.9913	1.0365

due to a scale gain, and 4.62% was VRS technical efficiency). As there was a drop in technological change in the order of 8.02% (1 – 0.9198), the technical efficiency gain of 8.25% resulted in a gain in the Malmquist Index of only 0.42%. From the second to the third period (2–3), there was a decrease in CRS technical efficiency of 1.6% (i.e. 1 – 0.984). However, there was a technological evolution in the sector of the order of 5.45%, resulting in an increase in total factor productivity (Malmquist Index) of 3.65%.

Project 2 reduced the technical efficiency and raised the technological change factor for the two grades under analysis. Thus, it is possible to infer that project 2 had better results in terms of efficiency than project 1.

10.4.7 STATISTICAL ANALYSES

Completing the longitudinal assessment of the efficiencies, the statistical analysis applied to the efficiency data aims to determine whether there is statistically significant variation in the means obtained. All statistical tests use the efficiency data with variable returns to scale (VRS), calculated considering a single period of analysis.

Chart 10.27 summarizes the significance (P value) obtained in the statistical analyses. It is observed that, in all tests performed, the significance was above 0.05. Such a result shows that the means of the efficiencies between the periods did not exhibit statistically significant differences. By means of these assessments, it is shown that

CHART 10.27

Significance of Applied Statistical Analyses

Test	Grade	P Period 1 and 2	P Period 2 and 3	P Period 1, 2 and 3
t	G1	0.458	0.466	...
	G2	0.291	0.938	
ANOVA	G1		...	0.482
	G2			0.495
Kruskal Wallis	G1		...	0.401
	G2			0.411

the impacts of the improvement and technological upgrade projects did not promote an increase in the efficiency of the process studied.

Thus, the statistical analysis of the efficiency variation between the periods shows that the implementation of improvement projects did not provide the positive effects expected in the company. Although small variations are observed in the efficiency indexes and Malmquist index, it is noticed that the variations identified are not statistically significant. This was shown from the t, ANOVA and Kruskal-Wallis tests.

10.5 EFFICIENCY AND CONTINUOUS IMPROVEMENT: CASE STUDY IN AN ARMS MANUFACTURING COMPANY

In this section, the results are presented in relation to the practical problem on productivity and efficiency management addressed in Section 3.5. First, we will briefly describe the working method adopted in the research, and, later, the results achieved.

10.5.1 WORKING METHOD

The research was developed in the Brazilian unit of a multinational armaments manufacturer. In this investigation, a longitudinal analysis was carried out considering the years 2007 to 2012. In the manufacturing process, each type of product is produced on a specific piece of equipment and with a dedicated set of tools. During the research, three types of products manufactured in three independent lines, that did not share resources, were assessed. The production lines were named C1, C2 and C3. The products are manufactured in monthly lots for each model. The monthly lots were considered as the DMUs of the research. The monthly lot consists of the total of products manufactured in the period of one month on each production line. Thus, a total of 216 DMUs was obtained. In addition, the use of the composite technical efficiency and the VRS model, with input orientation, were defined.

During the six-year analysis period, no machine was acquired to increase the capacity of the plant, since the production did not show significant increases in volume due to market restrictions. The major continuous improvements made in the period were in programs, such as A3 Reports, Kaizen events, and training programs for the production employees. Thus, data were collected on all continuous improvement projects, hours of employee training, production volumes and mean experience time employees spent on each production line. The following hypotheses are tested in the research:

H1a: There is a relationship between the use of A3 reports and the efficiency of the production process.

H1b: There is a relationship between the realization of Kaizen events and the efficiency of the production process.

H1c: There is a relationship between training of the production team and the efficiency of the production process.

H2: There is a relation between the experience time of the employees on the production line and efficiency of the production process.

H3a: There is a relationship between the use of A3 reports and the production volume.

H3b: There is a relationship between the realization of Kaizen events and the production volume.

H3c: There is a relationship between training of the production team and the production volume.

H4: There is a relationship between the experience time of the employees on the production line and the production volume.

H5: There is a relationship between production volume and efficiency.

For definition of the variables to be used in the DEA model, initially based on the literature, a pre-listing of variables was developed, later discussed and validated together with the specialists who supported the development of the study. Chart 10.28 presents the final listing of the variables used in the model, as well as their functions (input or output) and the unit of measurement used in the collection.

ANOVA and its assumptions (Shapiro-Wilk and Bartlett) and linear regression were also used to achieve a better understanding of the results.

10.5.2 DATA COLLECTION FROM THE PROJECTS FOR CONTINUOUS IMPROVEMENT, LEARNING AND PRODUCTION VOLUME

Information on the continuous improvement, learning and production volume projects, carried out over the period analyzed, was collected and segregated by production line (Chart 10.29). Among the continuous improvement projects, the total numbers of A3 reports, Kaizen events and training hours were collected. The learning analysis was limited to consider only the employee's experience, considering the mean length of stay on the production line. For the production volume, the total sum of pieces produced over the six years was considered.

CHART 10.28

Description and Detail of the Inputs and Outputs Used in the DEA Model

Variable	Theoretical Base	Function in the Model
Fabrication time	Jain et al. (2011); Park et al. (2014)	Input 1
Case	Jain et al. (2011); Park et al. (2014); Cook et al. (2014)	Input 2
Lead		Input 3
Powder		Input 4
Tools		Input 5
Primer		Input 6
Loaded cartridge	Jain et al. (2011); Cook et al. (2014)	Output 1
Empty cartridge	Jain et al. (2011); Cook et al. (2014)	Output 2

CHART 10.29
Quantity of Variables Used for Lines C1, C2 and C3 During the Period Analyzed

Relation	Variables studied	Line C1	Line C2	Line C3
Continuous	Number of A3 reports	5	10	5
improvement	Number of Kaizen events	9	42	38
	Hours of training	77.5	65.5	63.5
Learning	Time (years) operators spent on the line	6.06	5.67	4.97
Production volume	Parts produced per month	44,963,200	33,516,304	18,936,091

10.5.3 ANALYSIS OF RESULTS

After the data collection and treatment, the efficiency scores were calculated using DEA. Subsequently, the ANOVA test was performed. The ANOVA test aimed to determine whether there was a significant difference between the annual means of the efficiency scores, obtained through DEA. Chart 10.30 shows the mean efficiency for each year and the result of the ANOVA test.

The mean efficiency of Line C1 indicated low variability between years. The ANOVA test confirmed that the difference between the means of each year was not statistically significant ($P= 0.570$). The results indicate that the variables related to continuous improvement, learning and production volume did not have a significant effect on the increase of Line C1 efficiency over time. In this period, Line C1 was the one that had the least amount of Kaizen event type improvements (nine projects) and A3 reports (five projects), compared with C2 and C3. However, it was the one that had the highest mean in terms of the number of employees staying (6.06 years),

CHART 10.30
Analysis of Variance (ANOVA) Lines C1, C2 and C3

Year	Annual mean of DEA efficiency		
	Line C1	Line C2	Line C3
2007	0.5010	0.5123	0.4964
2008	0.5011	0.5047	0.5011
2009	0.4996	0.5010	0.5075
2010	0.5006	0.5032	0.5198
2011	0.5004	0.4983	0.5287
2012	0.5011	0.5048	0.5259
Standard Deviation	0.00232	0.00145	0.02003
F (ANOVA)	0.776	1.896	8.104
Significance (P-value)	0.570	0.107	0.000*

*P value < 0.05.

the highest number of training hours (77.5 hours) and the highest production volume (44,963,200 pieces). Furthermore, this is the line with newer equipment in relation to the other production lines.

For Line C2, the ANOVA test also confirmed that the difference between means was not significant (P value = 0.107). Line C2 had 42 Kaizen event projects and 10 A3 report projects during the period. Line C2 was the one that had the greatest number of improvement projects in relation to the others. In addition, 65.5 hours of training were carried out and the production volume was 33,516,304 pieces. Production Line C2 is the one with the equipment in a worse state of conservation, relative to Lines C1 and C3. The specialists pointed out that this is the line with the highest index of machine stops for maintenance, overtime and rework.

The mean of the composite technical efficiency of each year of Line C3 indicated greater variability relative to the others. The ANOVA test confirmed that the difference between the means of Line C3 was statistically significant ($P = 0.000$). This finding indicates improved efficiency over time. In Line C3, 38 Kaizen events projects were carried out, five A3 report projects and 63.5 training hours. However, Line C3 is the one with the shortest service time of employees (4.97 years) and the lowest production volume (18,936,091 pieces). According to the specialists, this is the line with the best state of conservation of equipment among the three surveyed.

Thus, based on the results presented, it was observed that only Line C3 improved efficiency over time, indicating better operational practices and positive influence of projects of continuous improvement, learning and production volume on efficiency. In addition, there were indications that the variables studied had no effect on the efficiency of Lines C1 and C2. However, to accept or refute the hypothesis of the research, the results concerning the multiple linear regression were analyzed. Based on this analysis, it was possible to test the hypothesis of the research.

10.5.4 Analysis of the Relationship among Continuous Improvement, Learning and Efficiency

Chart 10.31 presents the results of multiple linear regression with the analysis of the influence of independent variables related to continuous improvement and learning on the dependent variable efficiency for Lines C1, C2 and C3.

Lines C1 and C2 did not exhibit statistical significance in the ANOVA variance analysis of test, according to the result of the F test ($P = 0.857$, $P = 0.172$, respectively). In this case, the independent variables had no significant effects in terms of the efficiency of the two production lines. The results complement previous assessments. Thus, the results obtained in the multiple linear regression test indicate that the variables related to continuous improvement and learning had no significant influence ($P > 0.05$) on the result of the efficiency of Lines C1 and C2.

It can be observed, for example, that Line C2 was the one that had more Kaizen event projects and A3 reports, compared with Lines C1 and C3. However, the programs did not result in a significant improvement in efficiency. In summary, the data analyses show that there was no significant effect, in relation to the variables related to continuous improvement and learning, on efficiency of Lines C1 and C2.

CHART 10.31

Relationship among Continuous Improvement, Learning and Efficiency

Continuous improvement and learning – Independent variable	Efficiency – Dependent variable					
	Line C1		Line C2		Line C3	
	Standard beta	P	Standard beta	P	Standard beta	P
Number of A3 reports	−0.023	0.863	−0.074	0.537	−0.103	0.297
Number of Kaizen events	−0.011	0.936	0.151	0.220	0.337	0.010*
Hours of training	0.083	0.517	0.050	0.671	−0.043	0.677
Service time of the employees	0.091	0.477	0.211	0.092	0.440	0.000*
ANOVA – Teste F (p-value)	0.857		0.172		0.000*	
R	0.139		0.299		0.610	
R^2	0.019		0.090		0.373	
Adjusted R^2	−0.039		0.035		0.335	

*= P value < 0.05.

Line C3 presented statistical significance in the ANOVA test (P= 0.000). Standardized regression coefficients indicated that the employees' service time (β = 0.440, P value = 0.00) and the total number of Kaizen (β = 0.337, P value = 0.010) had a significant and positive influence on the efficiency of Line C3. The R^2 for Line C3 was 0.373, meaning that the sum of the independent variables explains 37.3% of the variance of the dependent variable (efficiency). It can be observed that Line C3, even with the shortest service time of employees (4.97 years), had a significant result in relation to the learning and efficiency increase. In view of the situation confirmed by the analysis, it is concluded that the variables employee service time (learning) and total number of Kaizen events (continuous improvement) had a significant influence on the efficiency of Line C3.

10.5.5 Analysis of the Relationship among Continuous Improvement, Learning and Production Volume

Chart 10.32 presents the results of multiple linear regression with the analysis of the influence of the independent variables related to continuous improvement and learning on the dependent variable, production volume, for the three production lines.

Line C1 presents statistical significance in the analysis of variance (P value = 0.056). For Line C1, the P value result was marginally above the threshold point. However, the result was considered to be significant, having, as a parameter the threshold point used in social research (P value $\leq$ 0.10). The R^2 score of Line C1 was 0.127, so the sum of the independent variables explains 12.7% of the variance of the dependent variable. The standardized regression coefficients indicated that the total

CHART 10.32

Relationship between Continuous Improvement, Learning and Production Volume

	Efficiency – Dependent variable					
Continuous improvement and learning – Independent variable	**Line C1**				**Line C1**	
	Standard beta	**P**	**Standard beta**	**P**	**Standard beta**	**P**
Number of A3 reports	−0.031	0.732	−0.044	0.717	0.038	0.740
Number of Kaizen events	−0.236	0.804	0.151	0.228	−0.278	0.022*
Hours of training	0.242	0.048*	−0.138	0.253	0.167	0.166
Service time of the employees	0.154	0.202	−0.151	0.233	−0.152	0.221
ANOVA – Test F (*P*-value)	0.056*		0.411		0.035*	
R	0.356		0.238		0.376	
R²	0.127		0.057		0.141	
Adjusted R²	0.075		0.000		0.090	

* = *P* value < 0.05.

number of hours of training had a positive and significant influence on the production volume of Line C1 ($\beta = 0.242$, $P = 0.048$). Line C2 did not result in statistical significance in the ANOVA (P value = 0.411), indicating that variables related to continuous improvement and learning had zero influence on the production volume.

Line C3 presents statistical significance in the ANOVA variance analysis (P value = 0.035). For Line C3, the R² was 0.141, indicating that the sum of the independent variables explains 14.1% of the variance of the dependent variable. For Line C3, the total number of Kaizen events had a significant negative influence on the production volume ($\beta = -0.278$, P value = 0.022). According to the process specialists, the volume increase in Line C3 results in a disproportionate consumption of tools and inputs. This fact suggests that production peaks compromise efficiency, overlapping the benefits obtained through Kaizen event projects. It is observed that the variable related to the service time of the employees (learning) was not significant on any of the three production lines.

10.5.6 ANALYSIS OF THE RELATIONSHIP BETWEEN EFFICIENCY AND PRODUCTION VOLUME

Chart 10.33 presents the results of simple linear regression with analysis of the influence of the independent variable, production volume, on the dependent variable, efficiency, for Lines C1, C2 and C3.

Lines C1 and C2 did not exhibit statistical significance in the ANOVA test (P value = 0.926 and P value = 0.808, respectively), indicating that production volume has no effect on the efficiency of these production lines. Line C3 presented statistical significance

CHART 10.33

Relationship between Efficiency and Production Volume

Production volume – Independent variable	Efficiency – Dependent variable					
	Line C1		Line C2		Line C3	
	Standard beta	P	Standard beta	P	Standard beta	P
Production volume	−0.011	0.926	−0.029	0.808	−0.448	0.000*
ANOVA F Test (Pvalue)	0.926		0.808		0.000*	
R	0.011		0.029		0.448	
R^2	0.000		0.001		0.200	
Adjusted R^2	−0.014		−0.013		0.189	

$* = P$ value < 0.05.

(P value $= 0.000$, $\beta = -0.448$). The standardized regression coefficients indicated that the production volume had a significant and negative relationship with the efficiency of Line C3. However, Line C3 presented an R^2 of 0.200, indicating that the independent variable (production volume) explains 20% of the variance of the dependent variable (efficiency). The study showed that Line C3 had the lowest production volume throughout the six years of analysis. However, it was the only line that had a significant influence of efficiency due to the increase in production volume. Therefore, it can be inferred that the increase in volume in Line C3 led to a reduction in efficiency.

An increase in the production volume without appropriate technological development can lead to productive inefficiency. There are indications that overloading on Line C3 contributed to increased waste and reduced efficiency. According to the specialists consulted, Line C3 did not have any process of technological upgrade during the analysis period. In addition, they claim to have observed the reduction in efficiency empirically as a function of volume increase. They also point out that the main inputs to be consumed in excess during the increase in volume are inputs 2 (case holster) and 3 (lead). One reason cited for this is the pressure and lack of time to achieve production goals imposed on the employees. As the volume increases, adjustments are neglected in order to avoid stopping the machines. This behavior increases the rate of tool breakage and the amount of scrap.

Another effect pointed out by the process specialists concerns the strength and durability of the Line C3 equipment. The increase in volume contributed to increasing problems related to machinery, such as wear of shafts, machine buses and devices. With these, the equipment suffers excessive play and vibration, making tools prone to breakage and the increased generation of scrap. Depending on the type of breakage, there may be misalignment of the machine bus, especially on double-effect type presses. These conditions contribute to the incidence of new operational problems. In addition to the wear problem, there is a risk of not stopping the equipment for maintenance or lubrication due to lack of time. This practice also contributes to breakdowns and waste.

The process specialists also highlighted employee satisfaction as an important aspect. With an excessive increase in the volume produced, the operational problems with machinery and equipment multiply. This effect leads to a high rate of overtime as well as to employee fatigue. In some cases, it generates notice requests, forcing the company to hire new operators. However, newly contracted operators require time to reach the level of productivity and experience required. Such a change also contributes to efficiency reduction. In addition, the fatigue process generated by operational problems and excessive overtime lead employees to negligence with regard to machinery adjustment, thus increasing tool breakage and waste generation.

10.5.7 SYNTHESIS OF RESULTS

The results obtained in the multiple linear regression test may respond to the tested hypotheses. The synthesis of the hypotheses tested is presented in Chart 10.34.

Regarding the relationship among continuous improvement, learning and efficiency, based on the results in Chart 10.34, the hypotheses H1a, H1b, H1c and H2 were refuted for the production Lines C1 and C2. This indicates that variables related to continuous improvement and learning had no relation to efficiency over time. The research hypothesis of the relationship among continuous improvement, learning and efficiency was partially supported in Line C3. Regarding continuous improvement, the number of Kaizen events, reflecting hypothesis H1b, was supported ($\beta = 0.337$, P value $= 0.010$). As for learning, the service time of H2 employees was supported ($\beta = 0.440$, P-value $= 0.00$). This indicates that the variables of Kaizen events and the service time of the employees on the production line helped to increase the efficiency of Line C3 over time.

As for the relation among continuous improvement, learning and production volume, the hypotheses were partially supported for Lines C1 and C3. In Line C1, the variable that showed a relation with the production volume was H3c, hours of training ($\beta = 0.242$, P value $= 0.048$). This indicates that the increase in this variable reflects an increase in the production volume. For Line C3, the variable that contributed to a negative relation of production volume was H3b, the number of Kaizen events ($\beta = -0.278$, P value $= 0.022$), indicating that the increase in volume reflects a reduction in the number of Kaizen events. This result presents indications that the increase in volume does not allow enough time for the realization of Kaizen projects. As for Line C2, all the hypotheses (H3a, H3b, H3c and H4) were refuted, indicating that the variables related to continuous improvement and learning had no relation with the Line C2 production volume.

The results referring to Lines C1 and C2 did not show a relationship between efficiency and production volume, thus refuting H5. The H5 research hypothesis was supported for Line C3 ($\beta = -0.448$, $R^2 = 0.200$, P value $= 0.000$). However, the relationship was negative, presenting a negative beta. The result indicates that the increase in the production volume leads to a reduction in the efficiency of Line C3. However, the relationship accounted for only 20% of the variance of the dependent variable.

CHART 10.34
Summary of the Results of the Hypotheses

Line	Research hypothesis		Standard beta	Sign. P-value	Test F P-value	R^2	Result of the hypotheses
C1	H1	H1a	−0.023	0.863	0.857	0,019	Refute
		H1b	−0.011	0.936			Refute
		H1c	0.083	0.517			Refute
	H2	H2	0.091	0.477			Refute
	H3	H3a	−0.031	0.732	0.056	0.127	Refute
		H3b	−0.236	0.804			Refute
		H3c	0.242	0.048*			Support
	H4	H4	0.154	0.202			Refute
	H5	H5	−0.011	0.926	0.926	0.000	Refute
C2	H1	H1a	−0.074	0.537	0.172	0.090	Refute
		H1b	0.151	0.220			Refute
		H1c	0.050	0.671			Refute
	H2	H2	0.211	0.092			Refute
	H3	H3a	−0.044	0.717	0.411	0.057	Refute
		H3b	0.151	0.228			Refute
		H3c	−0.138	0.253			Refute
	H4	H4	−0.151	0.233			Refute
	H5	H5	−0.029	0.808	0.808	0.001	Refute
C3	H1	H1a	−0.103	0.297	0.000*	0.373	Refute
		H1b	0.337	0.010*			Support
		H1c	−0.043	0.677			Refute
	H2	H2	0.440	0.000*			Support
	H3	H3a	0.038	0.740	0.035*	0.141	Refute
		H3b	−0.278	0.022*			Support
		H3c	0.167	0.166			Refute
	H4	H4	−0.152	0.221			Refute
	H5	H5	−0.448	0.000*	0.000*	0.200	Support

* = P value < 0.05.

BIBLIOGRAPHY

Cook, W., Tone, K., & Zhu, J. (2014). Data envelopment analysis: Prior to choosing a model. *Omega*, 44(1), 1–4.

Jain, S., Triantis, K.P., & Liu, S. (2011). Manufacturing performance measurement and target setting: a data envelopment analysis approach. *European Journal of Operational Research*, 214(3), 616–626.

Park, J., Lee, D., & Zhu, J. (2014). An integrated approach for ship block manufacturing process performance evaluation: Case from a Korean shipbuilding company. *International Journal of Production Economics*, 156(1), 214–222.

11 Concepts and Types of Modularity

Flávio I. Kubota and Paulo A. Cauchick-Miguel

This chapter seeks to present the concepts and types of product modularity. As modularization of products has been considered in the examples used in Chapters 3, 7 and 10, it is understood that an exclusive chapter on modularity can contribute to the reader's understanding of the subject. Thus, initially the conceptual differences concerning modularity and the modularization of products are presented. Subsequently, the concepts related to each type of modularity (product, production, use, organizational and service) are discussed. In addition, practical examples of the application of concepts are presented.

Before studying and expanding on the concept of modularity, a term focused on the result, or modularization, of a process-oriented concept to modularize products, processes or services, it is necessary to understand the concept of product architecture. Product architecture can be understood as a structure in which the functional components of the product are organized in physical parts and how the interactions occur among these through their interfaces (Ulrich, 1995).

The product architecture has two configurations: integrated and modular. An integrated architecture includes complex mapping of functional elements for physical components and non-standard coupled interfaces among components (Miltenburg, 2003; Piran et al., 2016).

In turn, modular product architecture is characterized by uncoupled interfaces and "one-to-one" mapping between functions and physical components. This allows each functional characterization element to be altered by changing its components, allowing the performance of each component to be independently modified (Baldwin & Clark, 1997; Lucarelli et al., 2015). Thus, modularity consists of a design approach that offers support to organizations through different technologies and decision-making processes, involving products and production processes (Lucarelli et al., 2015).

In this context, this chapter presents and explores the principles of modularity, applied to both products and services. The chapter begins with some general definitions of modularity in order to then present the different types of modularity, including some examples of each type. Finally, the last section of this chapter presents practical examples of modularity applications.

11.1   MODULARITY – GENERAL CONCEPTS AND DEFINITIONS

The classical definition of modularity states that, essentially, it consists of structuring a complex product or process through simpler subsystems, which can be designed independently, but which function or operate as a whole (Baldwin & Clark, 1997). The authors also point out that the IT sector has significantly increased its rate of innovation since the widespread adoption of modularity in its products and processes. In another definition, Ulrich and Eppinger (2012) regard modularity as the most important attribute of product architecture and as a way of organizing this architecture through parts that interact with each other. The term modularity is also used to describe the use of common units (or parts), having the ability to simply create variants of a product by changing some of its modules on a basic platform (Huang & Kusiak, 1998).

In general terms, the main objectives of modularity (Baldwin & Clark, 2000) are: (i) to facilitate the management of complex products and processes (as already mentioned at the beginning of this chapter), by subdividing them into simpler modules; (ii) to enable productive activities that are simultaneous and in parallel; and (iii) to adapt production to possible future uncertainties, since the final product can be modified by the adaptation of just one module, incurring considerably lower costs compared with completely remaking the product.

Thus, modularity, the types of which are described below, can increase efficiency in the product design and the manufacturing and assembly processes (Lugo-Márquez et al., 2016).

11.2   TYPES OF MODULARITY

There are different types of modularity, depending on the application focus. A typical classification includes product (design) modularity, production (process) modularity, which generally considers the manufacturing process, use modularity and organizational modularity. More recently, modularity has been applied to services, and this is currently an additional classification: service modularity. These types of modularity are detailed in sequence below.

11.2.1   MODULARITY OF PRODUCT

The modularity of design has some definition variants (Jacobs et al., 2011). Essentially, this type of modularity considers two main aspects of the product: its functionality (functions) and its physical structure (architecture). In the case of functionality, there is an association between the product's functions and the product's physical components, generating a sort of "map", and also considering the interfaces among the modules that comprise the product (Pandremenos et al., 2009).

In addition, product modularity consists of incorporating "blocks" that can be combined and recombined to provide significant diversity of product configurations (Baldwin & Clark, 1997; Jacobs et al., 2011). In this sense, product modularity features the use of standardized, interchangeable elements that allow a wide variety of end-products. Such a definition assumes high decoupling ability among modules, heterogeneous product configurations, and a function-to-module matching, practically one by one (Jacobs et al., 2011).

With regard to the types of product design information, there are two types that can guide a project: visible information (for example, internal and external norms) and invisible information (for example, those determined in the design process). The normative (visible) information is practically immutable, established in the initial phase of the project, and responsible for the integration of the final product (Carnevalli et al., 2011). In the understanding of the cited authors, this information defines not only each module, its interactions, functions and physical elements, but also its adjustments, connections and communications. In addition, the standards aim to ensure that each module is working properly, and also in relation to one another. On the other hand, the invisible information is that determined by the module designer, and the system architects do not need to be notified about possible modifications due to the independence among the modules (Graziadio, 2004).

The design modularity also facilitates generation of product families from a basic platform, through the combination of several modules. From this product variant, a high level of customization can be obtained (Chryssolouris et al., 2008). By adopting modularity, it is also possible to reuse modules and product parts (Paralikas et al., 2011).

The application possibilities of design modularity are varied, as defined by Ulrich (1995). Figure 11.1 presents an example of modularity application in product design, while Chart 11.1 presents five variations of the modularity concepts applied to products, followed by a brief explanation of these application variants.

With the possibility of sharing a single component or module across diverse products, this type of "modularity by sharing" can generate significant cost savings and

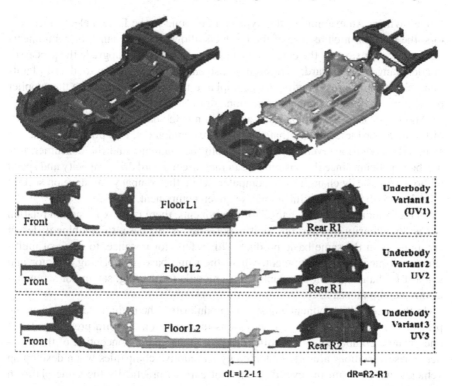

FIGURE 11.1 Example of application of modularity in design, comparing with an integral architecture. (From Paralikas et al., 2011)

CHART 11.1

Application Variations of Design/Product Modularity

Definitions and variations of the modularity concepts

Modularity by compartment	Same component shared among multiple products. Allows significant cost reduction by means of scope economies.
Modularity by component exchange	Complements modularity by sharing. This type is distinct, as different components may comprise the same basic product.
Modularity by component adjustment	Similar to modularity by component sharing and exchange. In this case, the dimensions of one or more components may vary within pre-established limits.
Modularity by mixture	Modularity that employs any one of the types above. However, in this concept, components are combined in such a way that they lose clear distinction among them.
Modularity by bus	Characterized by adoption of a base structure that can receive different types of components
Modularity by section	Modularity that provides the greatest degree of customization. Allows broad scope for configurations among different types of components, elements that constitute or form a particular product

Source: Ulrich (1995).

scope savings. An example of this type can be found in the IT and electronics sectors: the modular architecture of the products allows for the change of components separately; for example, the consumer can increase the RAM, upgrade the processor or add/change a video card, as desired. It also enables the customer to identify faults more clearly, and, in responding to a complaint, the company can trace the supplier responsible for the component in question (Agrawal et al., 2017).

Another possible example of this application is the hard disks (HDs). Regardless of the make or the company that manufactures computers, it is possible to use the same HD model for several makes. For both the company and the consumer, this can be beneficial, since the customers can increase his hard drive capacity and speed without necessarily changing the computer, while the company can develop several models with the same specifications as a given component.

In the "modularity by exchange" of components, there is an expansion in relation to the "modularity by sharing", as it is possible, in this case, to combine different components in the same basic product. This refers, for instance, to cases of multifunctional mechanical equipment, where the same base, comprising a motor and support for handling, allows coupling of a module for drilling, sawing or screwing/affixing components.

In the case of "component adjustment modularity", the main difference from the previous modes is that there are variations in dimensions (within preset limits) of one or more components, depending on the interest and/or conception of the product. These variations are often adopted by automotive companies when developing vehicles, which result in several variants of cars on practically the same platform (hatchback, sedan, etc.).

"Modularity by mix" involves the combination of the three previous types (component sharing, swapping and adjustment). However, in this case, there is a significantly varied combination, which brings about a loss of differentiation among the products developed.

Through "modularity by bus", it is possible to develop a product with a more diversified configuration, since a base structure is built to receive different components. Thus, the breadth of product variety increases and generates the potential to appeal to a more varied target audience. Finally, "sectional modularity" incorporates the other types of modularity applied to the product, approximating to the concept of mass customization.

It should also be noted that variations in the product modularity concept do not cancel out, but rather extend to one another, as a given variation is defined (from modularity through sharing to sectional modularity).

The modular product design is fundamental to facilitating the flexibility of organizations and businesses in general, since, from the initial design of the product, the layout of the industrial plant should be taken into account (Lugo-Márquez et al., 2016). In this sense, there are some studies that are more profound and which suggest analysis of the relationship between the application of modularity in product design and modularity in production processes. This latter type of modularity is detailed below.

11.2.2 MODULARITY OF PRODUCTION

This type of modularity, also called process modularity, deals with the ability to pre-combine a large number of elements (machines, workstations, etc.) into modules, which, later, can be assembled autonomously and independently (Pandremenos et al., 2009; Jacobs et al., 2011). Production modularity emphasizes the simplification of the assembly lines through pre-assemblies and functional pre-tests of the modules. This type can be introduced to fabricate products with three objectives (Persson, 2006): cost reduction, greater flexibility and minimization of the complexity of the tasks to be performed. It is also worth noting that some of the production activities can be transferred to suppliers (Arnheiter & Harren, 2005; Carnevalli et al., 2013).

The automotive sector is an example of the application of this type of modularity. Studies comparing traditional assembly and modular assembly (e.g. Baldwin & Clark, 1997; Van Hoek & Weken, 1998; Graziadio, 2004; Carnevalli et al., 2011) provide evidence of some of the principal characteristics of production modularity:

- Significant reduction of loss and of surplus components handled.
- Optimization of the transport of materials and inventory, which contributes to a reduction in the physical space required and a decrease in the taxation of transactions.
- Reduction of lead time, since it simplified and accelerated the production line.
- Decrease in production time, due to the parallel, rather than sequential, production of modules.

There are some distinct categories of production modularity commonly mentioned in the literature. Of these, two stand out: the industrial park and the modular consortium. Both concepts seek reductions in cost and development time. In the industrial park, first-tier suppliers construct facilities near the assembly plants, in some cases on the same site, from which deliveries of components or sub-assemblies occur in a just-in-time or sequenced just-in-time system (Dias, 1998). In this type of configuration, the author emphasizes that the main characteristic is the assembler in the role of leader, deciding which products will be supplied through the industrial park, which companies will supply these products, the location of these suppliers and how the deliveries will be made.

In the modular consortium, the first-level suppliers and the assembler operate "under one roof", the assembly of the vehicle being carried out by the suppliers or "partners"; the automaker does not have direct labor and the investments are shared (Cauchick Miguel & Pires, 2006). This model of production is considered "an industrial park taken to the extreme" (Dias, 1998, p. 85). The author also points out that, in the modular consortium, the suppliers are all located within the assembly plant of the assembler, and also perform the delivery, with the assembler focused on activities of engineering and product development, as well as other activities of management and decision-making (sales, marketing, customer service, etc.). When addressing production modularity, it becomes inevitable to mention the example of the modular consortium, MAN Latin America (MAN-LA;Volkswagen Trucks and Buses), located in Resende, RJ. This example is addressed in detail later in Section 11.3.

11.2.3 MODULARITY OF USE

In this type of modularity, the consumer is the one who defines some configurations of the product, characterizing a certain capacity of customization that can facilitate the use of the product by the consumers (Pandremenos et al., 2009). In addition, modularity of use can be used strategically to achieve differentiation, customization and an increase in the variety of products, as well as other advantages from an environmental and cost-reduction perspective (Kubota & Cauchick Miguel, 2013).

Through use modularity, the consumer has a series of options that allow a degree of personalization of the product so as to cater for individual interests (Lucarelli et al., 2015; Pine, 1993). In the automotive sector, this type of modularity is commonly seen as a configuration of restricted items by the consumer, although the modularity of use can go further and support innovations in terms of the customization of products, as well as offering possibilities to improve product maintenance (Lucarelli et al., 2015).

Still with respect to the automotive sector, an evident example of the application of this type of modularity is the Smart car, since the model is totally constituted by pre-assembled modules, some of which are illustrated in Figure 11.2. At the time of purchase, the customer has the possibility of choosing a customized variation of the vehicle at the dealership, and the factory then prepares it in a short time, according to the specifications, using modules in stock (Pandremenos et al., 2009; Piran, 2015).

FIGURE 11.2 Example of the application of Smart car modules. (Note: figure drawn by the authors. Main picture was taken by one of the authors and additional illustrations were taken from the Internet, e.g. Mercedes Benz.)

11.2.4 ORGANIZATIONAL MODULARITY

This type of application of modularity aims to reduce the complexity of activities involved in the management of the company or the production chain as a whole (Tsvetskova & Gustafsson, 2012). Thus, organizational modularity can contribute to potential changes in the management of complex systems, since a modular organization demands a high level of resource allocation capacity (Cheng, 2011; Piran, 2015).

The literature on organizational modularity can be divided into two distinct perspectives concerning the object of analysis (Campagnolo & Camuffo, 2009; Piran, 2015). The first focuses on investigating the relationship between product architectures and the organizational structure of the product development unit, while the second perspective focuses on the application of modularity throughout the organizational structure (Hoetker, 2006).

Thus, it is noted that one of the perspectives of organizational modularity is strongly related to production modularity, since it involves the relationship between the product architecture and the organizational structure of the company, which also involves decisions related to the outsourcing of modules and components, their definition by local or global suppliers, and the production layout of the industrial plant itself.

11.2.5 MODULARITY IN SERVICES

In addition to the types of modularity described above, a new approach has developed recently: modularity in services. One of the first works using this application was developed by Sundbo (1994), who considered modularization as a means to increase standardization in service provision. The application of modularity concepts has since been viewed as one of the great challenges in innovation and service design (Brax et al., 2017; Menor et al., 2002). Studies in the area of operations management have considered service supply based on processes (Brax, 2013; Brax et al., 2017), and dependent on consumer needs (Brax et al., 2017).

Moreover, it is to be stressed that the existing literature on service modularity used to be very focused on the development of product modules from a maintenance and service perspective (Gershenson et al., 2003), but, over time, it has developed towards functionalities and interdependencies among the service processes, seeking better treatment of the heterogeneity inherent in the tertiary sector (Geum et al., 2012; Pekkarinen & Ulkuniemi, 2006; Van Hoek & Weken, 1998). However, it is recognized that the modularization of services, in a practical context of application, is still an incipient theme and needs further study (Iman, 2016). There is also growing demand for knowledge about product architecture focused on the provision of services and business models in the service area (Brax et al., 2017).

11.3   EXAMPLES OF MODULARITY

This section describes some examples of modularity application, distinct from one another in terms of the types adopted, application purposes, and the results from the implementation of the modular strategy.

11.3.1   MODULAR CONSORTIUM – VOLKSWAGEN (MAN-LA)

Inaugurated in 1996 and located in Resende, 150 km from the city of Rio de Janeiro, this factory plant occupies about 1 million square meters. The Volkswagen modular consortium offers the domestic market about 20 models of trucks (between 8 and 42 tons) and five bus chassis, as well as exporting to more than 20 countries across Latin America, Central America and Africa (MAN-LA, 2017). The determination of the components of each truck module was defined by the product and manufacturing engineering teams of the company unit, with the following factors being considered in the decision-making (Ando, 2004; Carnevalli et al., 2011):

- Each module was considered to be a single unit, independent of others;
- The functions and parameters of each module;
- Product architecture, namely the functions of each module and their respective interfaces;
- Grouping by technological proximity (product and process) in each of the modules that make up the product, delegating the respective responsibilities of each module to a supplier with greater mastery of the technology.

Historically, the Modular Consortium has emerged from the demise of Autolatina, a joint venture led by Ford and Volkswagen (VW), from which VW needed to withdraw its bases to produce trucks and buses. From this, VW had the opportunity to design a brand new factory, based on the logic of modularity, with the explicit objective of reducing the total investment made by the automaker (Salerno & Dias, 1999), that is, to share this investment with partner suppliers. Thus, from the outset, the idea has been to focus only on activities with higher added-value from the automaker's point of view, such as engineering activities and product development.

In total, eight partners assemble complete sets of parts, subdivided into chassis (assembly), axles and suspension, wheels and tires, engines, cab frame, painting and cab finish. The monitoring and quality control of the products is the responsibility of MAN (created in 2009 from the acquisition of VW Trucks and Buses by MAN SE, the parent company of Grupo MAN), while the Modular Consortium seeks to reduce production costs, investment costs, inventories and production time, as well as to increase agility in the production of differentiated vehicles (MAN-LA, 2017). Figure 11.3 illustrates the activities and respective suppliers responsible. Figure 11.4 shows the initial layout after the implementation of the modular consortium.

This modular consortium production model is considered to be innovative in terms of management, and, over the years, has proven to be successful, leading to positive results. As previously mentioned, in this configuration, the company is able to dedicate itself to other important aspects of its business, such as logistics, marketing, customer service, and, mainly, development of new products (MAN-LA, 2017).

11.3.2 MODULAR COOKWARE – BLACK & DECKER®

Black & Decker is a North American company founded in 1910 in Baltimore, USA, which, since its foundation, has developed products aimed at the day-to-day life of

FIGURE 11.3 Activities and suppliers of the MAN Latin America modular consortium. (From MAN-LA, 2017.)

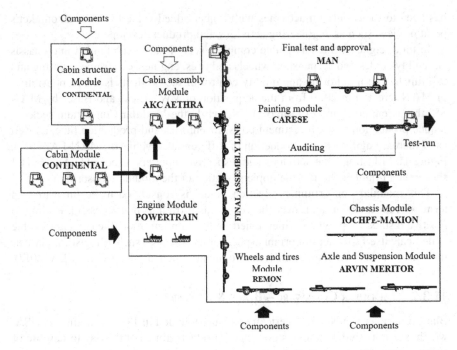

FIGURE 11.4   Assembly line of the modular consortium in Volkswagen. (From Pires, 1998.)

consumers all over the world. Thus, this company has developed products such as drills, electric irons and other domestic appliances. Of these products, in the context of modularity, the PE400 multifunctional cooker is outstanding, as it combines the functionalities of several electro-portable devices for customers (Black & Decker, 2017a). Figure 11.5 shows this product.

From the application of modularity, it was possible to design a multifunctional product, which serves several functions: steaming, frying, grilling and roasting. The product is subdivided into five main modules (Figure 11.5): glass lid, steam cooking module, grill module, anti-stick module and the heating base (Figure 11.6).

Thus, it is demonstrated that, in this case, the main benefit is the possibility of combining several functions within the same product, eliminating the need to purchase several products for each different function.

11.4   FINAL CONSIDERATIONS

The present chapter of this book addresses the main fundamentals of modularity, as well as their types and possible examples, where their application was demonstrated. Thus, it has been shown that modularity can be an important approach and also a way to seek innovation in products, processes, services and/or business models. Modularity approaches showing these results can be found in both the academic literature and in business/industry publications. However, modularizing products, processes and/or services is not a simple activity, as it demands a broad knowledge

FIGURE 11.5 Multifunctional cooker developed by Black & Decker®: application of modularity in design. (From Black & Decker, 2017b.)

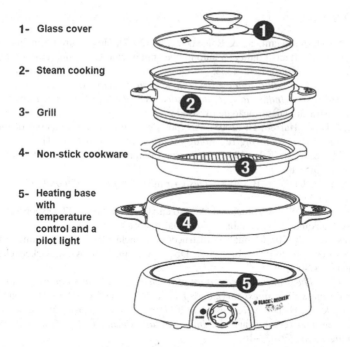

1- Glass cover

2- Steam cooking

3- Grill

4- Non-stick cookware

5- Heating base with temperature control and a pilot light

FIGURE 11.6 Multifunctional pan modules in PE400. (From Black & Decker, 2017b.)

of the product and the organizational processes, in addition to influencing, in some cases, the company's own organizational model.

In addition, one should consider the organization's own culture, since significant changes can occur due to the modularization of goods and services, as well as processes. Both the product design and the way the work is organized undergo alterations resulting from modularization. Thus, it is necessary to know the technical and cultural characteristics of the organization and, from this, develop modularity so that this can facilitate the exploration of new opportunities.

CHAPTER 11 QUESTIONS

It is recommended that the following issues be dealt with in order to aid assimilation of the contents of this chapter:

1. What is the general concept of modularity, and what are its main types?
2. Give two objectives of using modularity, and, for each goal, describe a practical example of how these goals can be achieved.
3. Discuss three possible benefits of modularity, and exemplify each with a practical case.
4. What is design modularity, and how can this typology be applied?
5. Discuss the main characteristics of the modularity explored in the case of the MAN Group.

REFERENCES

Agrawal, A., Muthulingam, S., & Rajapakshe, T. (2017). How sourcing of interdependent components affects quality in automotive supply chains. *Production and Operations Management*, 26(8), 1512–1533.

Ando, R. A. (2004). *Modularidade e seus Impactos no Desenvolvimento de Novos Produtos e Processos na Indústria Automobilística*. São Paulo, FEA USP, Departamento de Administração, Dissertação de Mestrado.

Arnheiter, E.D., & Harren, H. (2005). A typology to unleash the potential of modularity. *Journal of Manufacturing Technology Management*, 16(7), 699–711.

Baldwin, C.Y., & Clark, K.B. (1997). Managing in an age of modularity. *Harvard Business Review*, 75(5), 84–93.

Baldwin, C.Y., & Clark, K.B. (2000). *Design rules: The power of modularity*. Cambridge, MA: MIT Press.

Black & Decker. (2017a). Institucional. Acesso em: 01 dezembro, 2017. Disponível em: http://www.blackanddecker.com.br/institucional/historico.asp.

Black & Decker. (2017b). Manual de Instruções – Panela Elétrica Multi-Funcional PE400. Acesso em: 01 de dezembro, 2017. Disponível em: http://www.blackedecker.com.br/manual/PE400.pdf.

Brax, S. (2013). *The process based nature of services: Studies in management of industrial and business-to-business service*. Espoo: Aalto University, School of Science.

Brax, S., Bask, A., Hsuan, J., & Voss, C. (2017). Service modularity and architecture – an overview and research agenda. *International Journal of Operations & Production Management*, 37(6), 1–16.

Campagnolo, D., & Camuffo, A. (2009). What really drives the adoption of modular organizational forms? An institutional perspective from Italian industry-level data. *Industry and Innovation*, 16(3), 291–314.

Carnevalli, J.A., Cauchick Miguel, P.A., & Salerno, M.S. (2013). Aplicação da modularidade na indústria automobilística: análise a partir de um levantamento tipo survey. *Production*, 23(2), 329–344.

Carnevalli, J.A., Varandas Júnior, A., & Cauchick Miguel, P.A. (2011). Uma Investigação sobre os Benefícios e Dificuldades na Adoção da Modularidade em uma Montadora de Automóveis. *Produto & Produção*, 12(1), 60–90.

Cauchick Miguel, P.A., & Pires, S.R.I. (2006). A case study on modularity in product development and production within the auto industry. *International Journal of Automotive Technology and Management, 6*(3), 315–330.

Cheng, L.C. (2011). Assessing performance of utilizing organizational modularity to manage supply chains: Evidence in the US manufacturing sector. *International Journal of Production Economics, 131*(2), 736–746.

Chryssolouris, G., Papakostas, N., & Mavrikios, D. (2008). A perspective on manufacturing strategy: Produce more with less. *CIRP Journal of Manufacturing Science and Technology, 1*(1), 45–52.

Dias, A.V.C. (1998). *Consórcio modular e condomínio industrial: elementos para análise de novas configurações produtivas na indústria automobilística.* Tese (Mestrado em Engenharia de Produção) – Escola Politécnica da Universidade de São Paulo.

Gershenson, J.K., Prasad, G.J., & Zhang, Y. (2003). Product modularity: Definitions and benefits. *Journal of Engineering Design, 14*(3), 295–313.

Geum, Y., Kwak, R., Park, Y. (2012). Modularizing services: A modified HoQ approach. *Computers & Industrial Engineering, 62*(2), 579–590.

Graziadio, T. (2004). *Estudo Comparativo entre os Fornecedores de Componentes Automotivos de Plantas Convencionais e Modulares.* Tese (Doutorado em Engenharia de Produção), EPUSP, Departamento de Engenharia de Produção.

Hoetker, G. (2006). Do modular products lead to modular organizations? *Strategic Management Journal, 27*(1), 501–518.

Huang, C.C., & Kusiak, A. (1998). Modularity in design of products and systems. *IEEE Transactions on Systems, Management, and Cybernetics, 28*(1), 66–77.

Iman, N. (2016). Modularity matters: A critical review and synthesis of service modularity. *International Journal of Quality and Service Sciences, 8*(1), 38–52.

Jacobs, M., Droge, C., Vickery, S.K., & Calantone, R. (2011). Product and process modularity's effects on manufacturing agility and firm growth performance. *Journal of Product Innovation Management, 28*(1), 123–137.

Kubota, F.I., & Cauchick Miguel, P.A. (2013). Modularidade e desdobramento das função qualidade: uma análise teórica de publicações. *Revista Gestão Industrial, 9*(3), 700–726.

Lucarelli, M., Matt, D.T., & Russo Spena, P. (2015). Modular architectures for future alternative vehicles. *International Journal of Vehicle Design, 67*(4), 368–387.

Lugo-Márquez, S., Grisales, A.G., Rubio, O., & Eder, W.E. (2016). Modular redesign methodology for improving plant layout. *Journal of Engineering Design, 27*, 1–3.

MAN-LA. (2017). Institucional – Consórcio Modular. Acesso em: 07 nov. 2017. Disponível em: https://www.man-la.com/institucional/consorcio-modular.

Miltenburg, P.R. (2003). *Effects of modular sourcing on manufacturing flexibility in the automotive industry: A study among German OEMs.* Rotterdam: Erasmus University.

Menor, L.J., Tatikonda, M.V., & Sampson, S.E. (2002). New service development: Areas for exploitation and exploration. *Journal of Operations Management, 20*(2), 135–157.

Pandremenos, J., Paralikas, J., Salonitis, K., & Chryssolouris, G. (2009). Modularity concepts for the automotive industry: A critical review. *CIRP Journal of Manufacturing Science and Technology, 1*(3), 148–152.

Paralikas, J., Fysikopoulos, A., Pandremenos, J., & Chryssolouris, G. (2011). Product modularity and assembly systems: An automotive case study. *CIRP Annals – Manufacturing Technology, 60*(1), 165–168.

Pekkarinen, S., & Ulkuniemi, P. (2006). Modularity in developing business services by platform approach. *The International Journal of Logistics Management, 19*(1), 84–103.

Persson, M. (2006). Editorial introduction. *International Journal of Automotive Technology and Management, 6*(3), 257–261.

Pine, J.B. (1993). *Mass customization: The new frontier in business competition.* Boston (Massachusetts, EUA): Harvard Business School Press.

Piran, F.S. (2015). *Modularização de produto e os efeitos sobre a eficiência técnica: uma avaliação em uma fabricante de ônibus*. Dissertação (Mestrado em Engenharia de Produção e Sistemas). Universidade do Vale do Rio dos Sinos, São Leopoldo-RS.

Piran, F.S., Lacerda, D.P., Antunes Jr., J.A.V., Viero, C.F., & Dresch, A. (2016). Modularization strategy: Analysis of published articles on production and operations management (1999 to 2013). *International Journal of Advanced Manufacturing and Technology*, *86*(14), 507–519.

Pires, S. (1998). Managerial implications of the modular consortium model in a brazilian automotive plant. *International Journal of Operations & Production Management*, *18*(3), 221–232.

Salerno, M. S., & Dias, A. V. C. (1999). Product design modularity, modular production, modular organization: The evolution of modular concepts. *Automotive Industries*, *3*, 61–73.

Sundbo, J. (1994). Modulization of service production and a thesis of convergence between service and manufacturing organizations. *Scandinavian Journal of Management*, *10*(3), 245–266.

Tsvetskova, A., & Gustafsson, M. (2012). Business models for industrial ecosystems: A modular approach. *Journal of Cleaner Production*, 2930(7), 246–254.

Ulrich, K. (1995). The role of product architecture in the manufacturing firm. *Research Policy*, *24*(3), 419–440.

Ulrich, K., & Eppinger, S. (2012). *Product design and development* (5ª ed.). New York: McGraw-Hill.

Van Hoek, R.I., & Weken, H.A.M. (1998). The impact of modular production on the dynamics of supply chains. *The International Journal of Logistics Management*, *9*(2), 35–50.

12 Future Prospects

From the knowledge accumulated in this book, we conclude that new prospects will open up for the development of actions that could contribute to improvement of productivity and efficiency in Brazil and around the world. We have highlighted some of these actions. We understand that the DEA technique has been little applied by Brazilian and international companies. In this book, we have strived to show the potential of the technique, and how it can be useful to managers and companies. Furthermore, with the development of the MMDEA, we hope to facilitate the process of using DEA for analysis of productivity and efficiency by both managers and researchers.

Currently, mainly from the perspective of efficiency analysis in goods production systems, the technique widely used by Brazilian and international companies is OEE. However, our research, as described in this book, shows that OEE, by not considering all the elements, has limitations in assessment of the efficiency of a system. Thus, we highlight and demonstrate in practice that DEA can contribute to coping with the deficiencies OEE exhibits.

However, we believe that DEA need not replace the use of OEE, since both techniques can be used in an integrated manner. The example of integrated use that has been developed in this book is, as far as we know, the first proposal of its kind. There are many possibilities for applications and research on these topics.

Another future prospect, which we are working on at the moment, is to use the integrated DEA with process simulation, taking into account the modeling and dynamics of systems. DEA, in its conception, is used to assess efficiency, based on past data. However, it is possible to simulate efficiency by considering data from the simulation of different future scenarios. The data generated to calculate the efficiency with the use of DEA in future scenarios can be obtained from simulations with system dynamics models.

In addition, we understand that DEA may be used to assess the improvement effects made on the systems. As shown, one of these improvements lies in the modularity of products. Our research has shown that modularity contributes to increasing the productivity and efficiency of production systems. Modularity has also been little used in Brazil, and we believe that its use can and should be expanded. We intend to contribute in this regard.

Finally, we realize that the assessment of economic efficiency has barely scratched the surface. In both the academic and business contexts, there are many opportunities for analysis and assessment in this field. This is because a company can be technically efficient, but economically inefficient, depending on the prices of the inputs being considered in the model under assessment.

For example, consider assessment of the productivity and efficiency of two DMUs (A and B) with an input and an output, with DMU A using 1 kg of raw material to produce a product and DMU B, in turn, using 0.8 kg of raw material to produce the same amount of the same product. In this case, with respect to technical efficiency, DMU B is more productive (and consequently more efficient) because it uses less input to generate the same output.

Now, from the point of view of economic efficiency, let us say that the price paid per kg of raw material by DMU A is US$ 10.00. In this case, the cost of the product will be US$ 10.00 per kg (1 kg×US$ 10.00=US$ 10.00). However, as DMU B pays US$ 15.00 per kg of raw material, the cost will be US$ 12.00 (0.80 kg×US$ 15.00=US$ 12.00). Thus, DMU A is economically more productive (and consequently more efficient) than DMU B. Therefore, economic efficiency shows data relative to cash values and can more broadly represent the day-to-day realities of organizations.

Exercise List

1. A survey conducted by 'Exame' magazine in 2014 showed that the fixed-line telephone service in Brazil presented opportunities for growth. The report also highlighted that the companies invested heavily in the operational structure, and that there was an indication that the main route of increasing the results of the operators should be aimed at increasing the number of lines in operation. There are seven fixed-line operators operating in the Brazilian market, with different structures and scope of operations (with, for example, Sercomtel operating in PR State, whereas Embratel operates in all states).

   In this scenario, it is understood that it is important to assess the efficiency of the fixed-line segment in Brazil. It is, therefore, requested that, with the data available, a DEA model be developed, and then the efficiency be assessed, using the template provided with the variables and model available (Chart 12.1).

   Note: The answers for this exercise will be different according to the type of modeling performed.

CHART 12.1

Data from Brazilian Telephone Companies

Company	Number of lines in service	Network extension KM	Total no. of employees	Net operational revenue (US$)	Investment realized (US$)	Total no. of users with broadband Internet
ALGAR	895,677	24,120	2,237	743,600,000	155,000,000	369,587
EMBRATEL	11,794,000	63,000	7,664	9,800,000,000	3,500,000,000	3,713,245
GVT	4,339,219	54,000	17,387	4,800,000,000	2,200,000,000	2,727,644
NET	5,911,000	47,000	17,929	2,700,000,000	3,500,000,000	6,667,000
OI	17,380,043	1,183,456	14,684	10,300,000,000	722,000,000	5,258,000
SERCOMTEL	205,701	4,258	481	135,000,000	20,000,000	92,809
TELEFÔNICA	10,700,000	333,000	20,240	11,700,000,000	6,000,000,000	3,992,000

a) Which DEA (CRS/VRS) model was used and why?
b) Which orientation of the model was used and why?
c) Did the model present a discrimination problem? If so, why?
d) What was the Standard Efficiency of each DMU?
e) Using Composite Efficiency, which DMU was most efficient?
f) If the inefficiency of the inefficient DMU (with regard to Composite Efficiency) was treated, how many lines could be brought into service with the current quantity of resources?
g) How many lines could be added to the fixed-line telephone network in Brazil?
h) The most efficient DMU, serving as a benchmark, should be used with which DMU(s)?

2) The aim of the exercise is to carry out an assessment of the efficiency of the Brazilian port authorities in performing their functions, in the period 2007–2009, with use of DEA. The following are considered to be inputs: number of employees; operational cost; and investment. The cargo handled constitutes the output. Output orientation is suggested. The model chosen should be VRS, in view of the fact that no proportionality exists between inputs and outputs. Over a period of three years, eight port authorities were considered. The data available determine (Chart 12.2):
a) The standard efficiency, using the VRS model.
b) The efficiency of the inverted frontier.
c) Composite and standardized efficiency (identifying the most and least efficient DMUs).
d) Targets and slack of lower-performance DMU inputs.
e) Identify the two main DMUs of reference (benchmarks) for the sample.

3) The USA is the world's largest producer of electricity from wind power. However, it derives only 7% of its energy matrix from renewable sources, and is, therefore, highly dependent on polluting sources. Given this, it is necessary to assess the capacity for improvement in the US states' use of wind resources, seeking to diversify the country's energy matrix with an increase in the share of clean, renewable sources. For this assessment, the data-driven DEA technique is used with an output orientation to rank the US states with respect to efficiency in wind energy generation, including assessing the issue of accidents resulting from this production process. It is suggested to use the VRS model, although VRS and CRS are only considered to address point a). With the use of the data shown, the following should be determined, using the variable number of accidents as inputs (Chart 12.3):

Note: use the variable number of accidents as input (Chart 12.3).

a) The standard efficiency CRS (with constant returns to scale) and VRS (with variable returns to scale).
b) The efficiency of the inverted frontier.
c) Composite and normalized efficiencies (to identify the most and least efficient DMUs).
d) Targets and slack of lower-performance DMU inputs.
e) The two main DMUs for reference (benchmarks) for the sample.

CHART 12.2
Data from Brazilian Port Authorities

Port	Number of employees	Operational cost (thousand US$)	Investment (thousand US$)	Cargo handled (thousand tons)
A 2007	36	43,400	5	32,700
A 2008	36	40,284	7	33,500
A 2009	99	86,136	20	81,000
B 2007	54	66,843	29	61,300
B 2008	153	173,530	26	166,100
B 2009	171	266,480	62	257,300
C 2007	225	277,984	73	262,170
C 2008	315	410,933	62	401,127
C 2009	315	411,681	47	402,204
D 2007	261	335,231	50	324,198
D 2008	243	305,721	37	298,421
D 2009	270	394,032	36	388,400
E 2007	171	121,064	35	111,870
E 2008	288	303,092	38	297,900
E 2009	243	362,690	40	355,850
F 2007	324	390,503	64	381,740
F 2008	306	389,206	46	377,800
F 2009	324	370,785	68	362,550
G 2007	396	505,323	59	493,000
G 2008	333	438,085	61	426,718
G 2009	405	554,433	59	545,000
H 2007	378	494,391	51	488,561
H 2008	126	160,347	34	157,400
H 2009	126	159,495	23	156,600

4) A company producing war artifacts has a production line that manufactures land mines. For production of each batch of a particular type of mine, it was necessary to combine four types of explosives. Each batch produced different quantities of each explosive in different periods to obtain the same detonation power, according to military regulations. The efficiency of the company should be assessed, using DEA, considering each batch of mines produced to be a DMU. The orientation of the model for input is suggested. In addition, it is suggested to carry out two assessments, one with the CRS model and the other with the VRS model. With use of the data shown, the following should be determined (Chart 12.4):

a) The standard efficiency CCR (with constant returns to scale) and BCC (with variable returns to scale).

b) The efficiency of the inverted frontier.

c) The composite and normalized efficiency (to identify which batch exhibited higher or lower efficiency).

d) Targets and slack of the inputs of each batch of lower performance;

Identify the two main reference batches (benchmarks) in the sample.

CHART 12.3
Data from US Electricity Producers

State	Viable Area Available (thousand km²)	Installed Capacity (megawatts)	Number of Accidents	Generation (thousands megawatt-hours)
California	6.8	2,368	11	5,385
Colorado	77.4	1,063	2	3,221
Idaho	3.6	117	2	207
Illinois	50.0	962	5	2,337
Indiana	29.6	131	1	238
Iowa	114.1	2,635	3	4,084
Kansas	190.5	812	2	1,759
Maine	2.3	47	2	132
Massachusetts	0.2	2	1	4
Minnesota	97.9	146	11	4,355
Missouri	54.9	163	3	203
Montana	188.8	255	3	593
Nebraska	183.6	25	3	214
New Hampshire	0.4	24	1	10
New Jersey	0.0	8	2	21
New Mexico	98.4	496	1	1,643
New York	5.2	707	6	1,251
North Dakota	154.0	776	2	1,693
Ohio	11.0	7	2	15
Oklahoma	103.4	708	2	2,358
Wyoming	110.4	680	1	963

CHART 12.4
Data from Mines Companies

Lot	Explosive				Mines Produced
	PETN	TNT	ESTIFINATO	FULMINATO	
01	332	787	801	932	2781
02	198	395	416	465	2807
03	417	873	994	113	2,028
04	136	292	308	358	2,350
05	128	276	341	385	2,029
06	303	650	734	862	2,829
07	82	201	234	275	2,035
08	371	846	891	101	2,815
09	152	353	389	432	2,367
10	22	259	340	355	2,005
11	31	829	104	116	2,076
12	50	962	962	106	2,148
13	194	485	489	557	2,002
14	216	486	537	592	2,055
15	97	242	283	313	2,163

Index

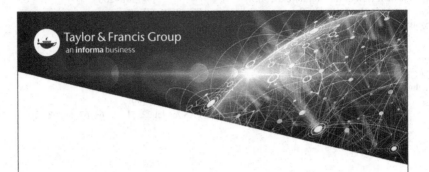

Printed in the United States
by Baker & Taylor Publisher Services